Advances in
Soil-borne Plant Diseases

Advances in Soil-borne Plant Diseases

Rob Jenkins
C.K. Jain

Oxford Book Company
Jaipur ▪ New Delhi

Published By

Oxford Book Company

Regd. Off. "Bony Residency", Gate No. 2
Opp. Tilak Public School, Vishweswariya Nagar,
Gopalpura Road, Jaipur - 302018, India
Ph: 0141 - 2761280, Telefax: 2761381
email - oxfordbookjpr@gmail.com
website - oxfordbookcompany.com

Delhi Office: 102, 1st Floor, Satyam House,
4327/3, Ansari Road, Darya Ganj,
New Delhi - 110002, India
Ph: 011-45652440

Distributed by
KEYA INTERNATIONAL PRESS LLP
E-mail: keyainternationalpressllp@gmail.com

Published in India, 2025

ISBN : 9789380179155

Typeset by Shivangi Computers, Jaipur

Printed and bou[illegible]y Replika Press Pvt. Ltd.

Preface

Plant diseases result when a susceptible host and a disease-causing pathogen meet in a favourable environment. If any one of these three conditions were not met, there would be no disease. Diseases may occur in natural environments, but they rarely run rampant and cause major problems. In contrast, the threat of disease epidemics in crop production is constant. The reasons for this are becoming increasingly evident. Soil-borne diseases result from a reduction of biodiversity of soil organisms. Restoring beneficial organisms that attack, repel, or otherwise antagonise disease-causing pathogens will render a soil disease-suppressive. Plants growing in disease-suppressive soil resist diseases much better than in soils low in biological diversity. Beneficial organisms can be added directly, or the soil environment can be made more favourable for them through use of compost and other organic amendments. Compost quality determines its effectiveness at suppressing soil-borne plant diseases.

There are two types of disease suppression: specific and general. Specific suppression results from one organism directly suppressing a known pathogen. These are cases where a biological control agent is introduced into the soil for the specific purpose of reducing disease incidence. General suppression is the result of a high biodiversity of microbial populations that creates conditions unfavourable for plant disease development. Soil pH, calcium level, nitrogen form, and the availability of nutrients can all play major roles in disease management. Adequate crop nutrition makes plants more tolerant of or resistant to disease. Also, the nutrient status of the soil and the use of particular fertilisers and amendments can have significant impacts on the pathogen's environment.

This book describes the characteristics of various soil-borne diseases of plants and the measures to prevent them. It provides a detailed description of the life of the pathogens, chemical and pesticidal means of regulation, taxonomic changes have been made in bacteria, fungi, nematodes and viruses; changing patterns of diseases and many recently reported diseases. This book should be useful to gardeners, landscape architects, florists, nurserymen, seed and fungicide dealers, pesticide applicators, cooperative extension agents and plant pathologists. It should also be a u[illegible] reference book for plant pathology classrooms and in some cases used as a te[illegible]

Rob Jenkins
C.K. Jain

Contents

1

Introduction to Soil-borne Plant Diseases

Soil-borne diseases result from a reduction of biodiversity of soil organisms. Restoring beneficial organisms that attack, repel, or otherwise antagonize disease-causing pathogens will render a soil disease-suppressive. Plants growing in disease-suppressive soil resist diseases much better than in soils low in biological diversity. Beneficial organisms can be added directly, or the soil environment can be made more favorable for them through use of compost and other organic amendments. Compost quality determines its effectiveness at suppressing soil-borne plant diseases. Compost quality can be determined through laboratory testing.

Plants can suffer from bacterial, viral and fungal attack just as we can ourselves. The organisms themselves (pathogens) are different, but at the microbial level the infection is much the same since one cell is as good a host as another. No matter which part of the plant is attacked the effect is usually to weaken or kill it. By infecting the leaves the plant's ability to produce its food is reduced. Some pathogens block the vessels in the stems which supply the leaves and by attacking the roots, the uptake of water and nutrients is reduced or stopped completely.When a plant is attacked by one of these microorganisms the damage caused provides an opportunity for the others to get in and it is the combined onslaught which deals the final blow. Also if it is under stress, such as through drought or poor nutrition it is more susceptible.

Sometimes the 'infection' is symbiotic where both organisms derive a benifit. A good example of this is the nitrogen fixing bacteria (Rhizobium) which reside in nodules on the roots of leguminous (pea family) plants, the plant provides food and protection, the bacteria takes nitrogen from the air and converts it to a form usable by the host. Also the Mycorrhizae are a whole Order of fungi which have a symbiotic relationship with plant roots. In other cases of interest to gardeners, the plant is not benefited but the changes caused produce more attractive features. This is what happens with Aucuba japonica where a viral infection produces the mottled leaves in the 'Variegata' variety. The Tulipomania craze in the 17th century was caused when 'Broken' varieties of tulips began to appear with streaked and mottled petals. This occurred at random and increased the desire and fascination for the bizarre effects. Prized examples were valued at more than "a mansion with servants", although it was not discovered until the early 20th century that viruses were to blame for the unusual colours and effects. In Holland, where the craze caused great hardship when it crashed in 1637, they now grow the pure forms with no streaks or frills.

Fungi are essential in breaking down dead organic matter to produce the humus which is needed for good soil structure—saprophytes. They do not have any chlorophyll so cannot use light to capture energy, instead they derive their energy by breaking down plant and animal material—alive or dead. They can also live in a symbiotic relationship, eg. the micorrhiza in the fine roots of conifers which cannot survive without them to take up vital nutrients. The widespread use of chemical control can damage the balance of these beneficial fungi and this forms part of the principals of Organic management. Lichens are an algae and a fungus growing together as a symbiotic conjunction, i.e., the fungus provides physical support for the algae and the algae produces food. There are some less welcome fungi which attack living plants and weaken or kill them—these are the ones which are mentioned in more detail here.

Viruses dwell inside the cells and cannot be treated with chemicals so affected plants must be destroyed (special microculture techniques may overcome the infection by taking cells from the growing tip, but this is restricted to the laboratory). There are no antibiotics for plants, so bacterial attacks, eg. fireblight, are untreatable as well. Fungi can be killed with chemicals without damaging the host because their growth habit is different, ie. they tend to grow on the plant and not in it, using root-like structures to extract nurishment.

Since killing the pathogens is difficult or impossible, "prevention is better than cure". By observing good hygiene when propogating and growing your plants, you can prevent a lot of diseases from taking hold.

— Destroy diseased plants, clear up dead leaves and other debris.

— As the spores can have tough outer coatings, do not add diseased material to the domestic compost heap as they do not usually achieve high enough temperatures to destroy them.

— Prune fruit trees and bushes regularly to keep an open structure allowing a good air flow and to remove damaged branches.

— Disinfect secateurs, saws or knives used for cutting out diseased branches with methylated spirits or a flame (a cigarette lighter comes in handy for this). Household bleach or other disinfectants can also be used as a dip for shears and clippers, including electric or petrol machines. This also helps when taking cuttings.

- Only use new or well-washed containers when growing cuttings and sowing seeds.
- Crop rotation in the vegetable plot will prevent a build up of disease.
- Space plants well apart especially crops where similar plants are growing together, to allow good air flow. Fungal diseases in particular, thrive in still, damp air and there is a greater chance for them to be transmitted to surrounding plants if they are in close proximity.
- Catching disease early is important so keep an eye out for it at all times.
- Plants are more susceptible to disease if they are not growing well. This can be due to poor soil, drought or both. So prepare the site well adding plenty of organic matter and give plants an occasional feed. As gardeners we are usually trying to grow plants which most likely are not native to the region or in a place they would not choose for themselves, so there is a greater probability that they could be under stress.

TYPES OF SOIL-BORNE PLANT DISEASES

Plant diseases result when a susceptible host and a disease-causing pathogen meet in a favorable environment. If any one of these three conditions were not met, there would be no disease. Many intervention practices (fungicides, methyl bromide fumigants, etc.) focus on taking out the pathogen after its effects become apparent. This publication emphasizes making the environment less disease-favorable and the host plant less susceptible.

Plant diseases may occur in natural environments, but they rarely run rampant and cause major problems. In contrast, the threat of disease epidemics in crop production is constant. The reasons for this are becoming increasingly evident. Dr. Elaine Ingham, a soil microbiologist and founder of Soil Foodweb Inc., describes the progression from undisturbed grassland—where a wide diversity of plants grow, their roots commingling with a wide diversity of soil organisms—to a field in row crops.

A typical teaspoon of native grassland soil would contain between 600- to 800-million individual bacteria that are members of perhaps 10,000 species. There are several miles of fungi, and perhaps 5000 species of fungi per teaspoon of soil. There are 10,000 individual protozoa split into three main groups, i.e., flagellates, amoebae and ciliates, and perhaps 1000 species of protozoa. There are 20 to 30 beneficial nematodes, which are members of as many as 100 species. Root-feeding nematodes are quite scarce in truly healthy soils. They are present, but in numbers so low that it is rare to find them. After only one plowing a few species of bacteria and fungi become extinct locally because the food they need is no longer put back in the system. But for the most part, all the suppressive organisms, all the nutrient cyclers, all the decomposers, all the soil organisms that rebuild good soil structure are still present and continue to try to do their jobs.

Why doesn't the limited food resources bother them more? A good savings account of organic matter has been built up in native grassland and native forest soil. The soil organisms use the organic matter they "put away" all those years when disturbance did not occur. ...But agriculture continues to mine soil organic matter and kill fungi by tilling. The larger predators are crushed, their homes destroyed. The bacteria go through a bloom and blow off huge amounts of that savings account organic-matter. With continued tillage the "policemen" (organisms) that compete with and inhibit disease are lost. The "architects" that build soil aggregates, are lost. So are the engineers, the larger organisms that design and form the larger pores in soil. The predators that keep bacteria, fungi and root-feeding organisms in line are lost. Disease suppression declines, soil structure erodes, and water infiltration decreases because mineral crusts form.

The decline can take 20 to 30 years to reach the point that most of the natural controls are finally lost and disease runs rampant. The speed with which the "edge" is reached depends on the amount of soil organic matter that was in the soil when it was first plowed, how often the soil was plowed and how much residue was added back. Additionally, how much variety was added back, and the inoculum base for the disease are also important. Certain diseases don't occur in some places because the disease hasn't reached them yet. But the instant the disease does arrive, it goes throughout he fields like a wildfire, because there are few natural competitors to stop it in the soil.

This progression of decline that Dr. Ingham describes leads to sick soils, and sick soils produce sick crops. As plants and soils have become sicker, growers have responded with newer and more powerful chemicals in an effort to kill off the problem pathogens. While it may seem the logical course of action, chemical intervention only serves to make things worse over time. Many pesticides reduce the diversity of soil life even further and select for resistant pathogens. This is the history of methyl bromide. Once this fumigant was highly effective if used only every five years. Today, on the same soils, it must be used much more frequently to keep the pathogens under control.

Until we improve the soil life we will continue on this pesticide treadmill. The general principle is to add the beneficial soil organisms and the food they need—the ultimate goal being the highest number and diversity of soil organisms. The higher the diversity, the more stable the soil biological system. These beneficial organisms will suppress disease through competition, antagonism, and direct feeding on pathogenic fungi, bacteria, and nematodes. We cannot restore the balance of organisms that was present under native, undisturbed circumstances, but we can build a new, stable balance of soil organisms that will be adapted to the altered soil conditions. This is a proactive plan that moves us toward the desired outcome of disease prevention.

STRATEGIES FOR CONTROL

There are two types of disease suppression: specific and general. Specific suppression results from one organism directly suppressing a known pathogen. These are cases where a biological control agent is introduced into the soil for the specific purpose of reducing disease incidence. General suppression is the result of a high biodiversity of microbial populations that creates conditions unfavorable for plant disease development. A good example of specific suppression is provided by a strategy used to control one of the organisms that cause damping off—Rhizoctonia solani. Where present under cool temperatures and wet soil conditions, Rhizoctonia kills young seedlings. The beneficial fungus Trichoderma locates then attacks Rhizoctonia through a chemical released by the pathogen. Beneficial fungal strands (hyphae) entangle the pathogen and release enzymes that dehydrate Rhizoctonia cells, eventually killing them. Introducing a single organism to soils seldom achieves disease suppression for very long. If not already present, the new organism may not be competitive with existing microorganisms. If food sources are not abundant enough, the new organism will not have enough to eat. If soil conditions are inadequate, the introduced beneficial organism will not survive. This practice is not sufficient to render the soil "disease suppressive"; it is like planting flowers in the desert and expecting them to survive without water. With adequate soil conditions, inoculation with certain beneficials should only be needed once.

Disease Suppressive Soils

A soil is considered suppressive when, in spite of favorable conditions for disease to occur, a pathogen either cannot become established, establishes but produces no disease, or establishes and produces disease for a short time and then declines. Suppressiveness is linked to the types and numbers of soil organisms, fertility level, and nature of the soil itself (drainage and texture). The mechanisms by which disease organisms are suppressed in these soils include induced resistance, direct parasitism (one organism consuming another), nutrient competition, and direct inhibition through antibiotics secreted by beneficial organisms. Additionally, the response of plants growing in the soil contributes to suppressiveness. This is known as "induced resistance" and occurs when the rhizosphere (soil around plant roots) is inoculated with a weakly virulent pathogen. After being challenged by the weak pathogen, the plant develops the capacity for future effective response to a more virulent pathogen. In most cases, adding mature compost to a soil induces disease resistance in many plants.

The level of disease suppressiveness is typically related to the level of total microbiological activity in a soil. The larger the active microbial biomass, the greater the soil's capacity to use carbon, nutrients, and energy, thus lowering their availability to pathogens. In other words, competition for mineral nutrients is high, as most soil nutrients are tied up in microbial bodies. Nutrient release is a consequence of grazing by protozoa and other microbial predators: once bacteria are digested by the predators, nutrients are released in their waste.

High competition—coupled with secretion of antibiotics by some beneficial organisms and direct parasitism by others—makes a tough environment for the pathogen. Our goal is to create soil conditions with all three of these factors present. Therefore, we want high numbers and diversity of competitors, inhibitors, and predators of disease organisms, as well as food sources on which these organisms depend. The food for beneficial organisms comes either directly or indirectly from organic matter and waste products from the growth of other organisms.

Limiting available nutrients is a key for general suppression. With an abundance of free nutrients, the pathogen can prosper. Virtually any treatment to increase the total microbial activity in the soil will enhance general suppression of pathogens by increasing competition for nutrients. So, how does the plant survive without readily available nutrients? It does so through microbial associations with mychorrhizal fungi and bacteria that live on and near the roots. These microbes scavenge nutrients for the plant to use. In return the plant provides carbon in the form of sugars and proteins to the microbes. This symbiotic system supports the beneficial organisms and the plant, but generally excludes the pathogens that would attack the plant.

It should be noted that general suppression will not control all soil-borne diseases. Rhizoctonia solani and Sclerotium rolfsii, for example, are not controlled by suppressive soils—their large propagules make them less reliant on external energy or nutrient sources, and therefore, they are not susceptible to microbial competition. With these two pathogens, "specific" beneficial organisms such as Trichoderma and Gliocladium will colonize the harmful propagules and reduce the disease potential.

Mycorrhizal Fungi and Disease Suppression

Among the most beneficial root-inhabiting organisms, mycorrhizal fungi can cover plant roots, forming what is known as a fungal mat, The mycorrhizal fungi protect plant roots from diseases in several ways:

- By providing a physical barrier to the invading pathogen. A few examples of physical exclusion have been reported. Physical protection is more likely to exclude soil insects and nematodes than bacteria or fungi. However, some studies have shown that nematodes can penetrate the fungal mat.
- By providing antagonistic chemicals. Mycorrhizal fungi can produce a variety of antibiotics and other toxins that act against pathogenic organisms.
- By competing with the pathogen.

- By increasing the nutrient-uptake ability of plant roots. For example, improved phosphorus uptake in the host plant has commonly been associated with mychorrhizal fungi. When plants are not deprived of nutrients, they are better able to tolerate or resist disease-causing organisms.
- By changing the amount and type of plant root exudates. Pathogens dependent on certain exudates will be at a disadvantage as the exudates change.

In field studies with eggplant, fruit numbers went from an average of 3.5 per plant to an average of 5.8 per plant when inoculated with Gigaspora margarita mycorrhizal fungi. Average fruit weight per plant went from 258 grams to 437 grams. A lower incidence of Verticillium wilt was also realized in the mycorrhizal plants.

Protection from the pathogen Fusarium oxysporum was shown in a field study using a cool-season annual grass and mycorrhizal fungi. In this study the disease was suppressed in mycorrhizae-colonized grass inoculated with the pathogen. In the absence of disease the benefit to the plant from the mycorrhizal fungi was negligible. Roots were twice as long where they had grown in the presence of both the pathogen and the mycorrhizal fungi as opposed to growing with the pathogen alone. Great care was taken in this study to assure that naturally-occurring mycorrhizal species were used that normally occur in the field with this grass, and that their density on the plant roots was typical.

Crop Rotation and Disease Suppression

Avoiding disease buildup is probably the most widely emphasized benefit of crop rotation in vegetable production. Many diseases build up in the soil when the same crop is grown in the same field year after year. Rotation to a non-susceptible crop can help break this cycle by reducing pathogen levels. To be effective, rotations must be carefully planned. Since diseases usually attack plants related to each other, it is helpful to group vegetable rotations by family—e.g., nightshades, alliums, cole crops, cucurbits. The susceptible crop, related plants, and alternate host plants for the disease must be kept out of the field during the rotation period. Since plant pathogens persist in the soil for different lengths of time, the length of the rotation will vary with the disease being managed. To effectively plan a crop rotation, it is essential to know what crops are affected by what disease organisms.

In most cases, crop rotation effectively controls those pathogens that survive in soil or on crop residue. Crop rotation will not help control diseases that are wind-blown or insect vectored from outside the area. Nor will it help control pathogens that can survive long periods in the soil without a host—Fusarium, for example. Rotation, by itself, is only effective on pathogens that can overwinter in the field or be introduced on infected seeds or transplants. Of course, disease-free transplants or seed should be used in combination with crop rotation. The period of time between susceptible crops is highly variable, depending on the disease. For example, it takes seven years without any cruciferous crops for clubfoot to dissipate. Three years between parsley is needed to avoid damping off, and three years without tomatoes to avoid Verticillium wilt on potatoes.

A three-year crop rotation is the standard recommendation for control of black rot (Ceratocystis fimbriata), stem rot (Fusarium oxysporum), and scurf (Monilochaetes infuscans) in sweet potatoes. Rotations may include grasses, corn, and other cereals in the Southwest where Texas root rot (Phymatotrichum omnivorum) is a problem.

Table 1. Rotation periods to reduce vegetable soil-borne diseases.

Vegetable	Disease	Years w/o susceptible crop
Asparagus	Fusarium rot	8
Beans	Root rots	3–4
Cabbage	Clubroot	7
Cabbage	Blackleg	3–4
Cabbage	Black rot	2–3
Muskmelon	Fusarium wilt	5
Parsnip	Root canker	2
Peas	Root rots	3–4
Peas	Fusarium wilt	5
Pumpkin	Black rot	2
Radish	Clubroot	7

Plant Nutrients and Disease Control

Soil pH, calcium level, nitrogen form, and the availability of nutrients can all play major roles in disease management. Adequate crop nutrition makes plants more tolerant of or resistant to disease. Also, the nutrient status of the soil and the use of particular fertilizers and amendments can have significant impacts on the pathogen's environment.

One of the most widely recognized associations between fertility management and a crop disease is the effect of soil pH on potato scab. Potato scab is more severe in soils with pH levels above 5.2. Below 5.2 the disease is generally suppressed. Sulfur and ammonium sources of nitrogen acidify the soil, also reducing the incidence and severity of potato scab. Liming, on the other hand, increases disease severity. While lowering the pH is an effective strategy for potato scab, increasing soil pH or calcium levels may be beneficial for disease management in many other crops. Adequate levels of calcium can reduce clubroot in crucifer crops (broccoli, cabbage, turnips, etc.). The disease is inhibited in neutral to slightly alkaline soils (pH 6.7 to 7.2). A direct correlation between adequate calcium levels, and/or higher pH, and decreasing levels of Fusarium occurrence has been established for a number of crops, including tomatoes, cotton, melons, and several ornamentals. Soil pH, calcium level, nitrogen form, and the availability of nutrients can all play major roles in disease management.

Calcium has also been used to control soil-borne diseases caused by Pythium, such as damping off. Crops where this has proved effective include wheat, peanuts, peas, soybeans, peppers, sugarbeets, beans, tomatoes, onions, and snapdragons. Researchers in Hawaii reported reduction of damping off in cucumber after amending the soil with calcium and adding alfalfa meal to increase the microbial populations.

Nitrate forms of nitrogen fertilizer may suppress Fusarium wilt of tomato, while the ammonia form increases disease severity. The nitrate form tends to make the root zone less acidic. Basically, the beneficial effects of high pH are lost by using acidifying ammonium nitrogen. Tomato studies have shown that use of nitrate nitrogen in soil with an already high pH results in even better wilt control. Celery studies showed reduced Fusarium disease levels from using calcium nitrate as compared to ammonium nitrate. The nitrate nitrogen form also produced the lowest levels of Fusarium on chrysanthemums, king aster, and carnation.

It has long been known that the form of nitrogen fertilizer can influence plant disease incidence. Research is beginning to reveal why. Dr. Joe Heckman of Rutgers University showed that when grass roots absorbed nitrate nitrogen, an alkaline root zone condition was created. When the grass absorbed ammonium nitrogen, an acid root zone was created. The pathogen responsible for summer patch disease in turf thrives in alkaline soils. This finding supported the use of ammonium sulfate for grass. Research trials using ammonium sulfate reduced summer patch severity up to 75%, compared to using an equal rate of calcium nitrate. A more acid soil also fosters better uptake of manganese. Adequate manganese stimulated disease resistance in some plants. Research at Purdue University showed that uptake of ammonium nitrogen improved plant uptake of manganese and decreased take-all disease (Gaeumannomyces graminis var. tritici). Similar results were seen with Verticillium wilt in potatoes and stalk rot in corn.

Potassium fertility is also associated with disease management. Inadequate potash levels can lead to susceptibility to Verticillium wilt in cotton. Mississippi researchers found that cotton soils with 200 to 300 pounds of potassium per acre grew plants with 22 to 62% leaf infections. Soil test levels above 300 pounds per acre had from zero to 30% infection rate. High potassium levels also retard Fusarium in tomatoes. Severity of wilt in cotton was decreased by boosting potassium rates as well.

Phosphate can also be critical. Increasing phosphorus rates above the level needed to grow the crop can increase the severity of Fusarium wilt in cotton and muskmelon. In general, the combination of lime, nitrate nitrogen, and low phosphorus is effective in reducing the severity of Fusarium.

Compost and Disease Suppression

Compost has been used effectively in the nursery industry, in high-value crops, and in potting soil mixtures for control of root rot diseases. Adding compost to soil may be viewed as one of a spectrum of techniques—including cover cropping, crop rotations, mulching, and manuring—that add organic matter to the soil. The major difference between compost-amended soil and the other techniques is that organic matter in compost is already "digested." Other techniques require the digestion to take place in the soil, which allows for both anaerobic and aerobic decomposition of organic matter. Properly composted organic matter is digested chiefly through aerobic processes. These differences have important implications for soil and nutrient management, as well as plant health and pest management. Chemicals left after anaerobic decomposition largely reduce compost quality. Residual sulfides are a classic example.

Successful disease suppression by compost has been less frequent in soils than in potting mixes. This is probably why there has been much more research (and commercialization) concerning compost-amended potting mixes and growing media for greenhouse plant production than research on compost-amended soils for field crop production. Below is a table that outlines some of the (mostly) field research done on compost-amended soils and the effects on plant disease.

In some further research, University of Florida field trials showed disease suppressive effects of compost and heat-treated sewage sludge on snap beans and southern peas (black-eyed peas). The compost was applied at 36 or 72 tons per acre and the sludge at 0.67 and 1.33 tons per acre. Bush beans were planted six weeks after the organic treatments were applied and tilled in. After the bush beans were harvested, a second crop of southern peas was planted. A standard fertilizer program was used. Plant damage from ashy stem blight was given a rating of slight, moderate, or severe. Rhizoctonia root rot disease ratings were made using a scale from 0 to 10, where 10 represented the most severe symptoms.

Bean sizes from the compost treatment, at both application rates (36 and 72 T/ac), were larger and yields 25% higher than those from areas receiving no organic amendment. Ashy stem blight was severe in areas with no compost applied. The disease was reduced under the sludge treatment but almost eliminated where compost had been applied. Leaf wilting and leaf death were pronounced in that portion of the field where compost was not applied.

Southern peas as a second crop had greener foliage and larger plants under both ates of compost. Pea yields were significantly higher with 36 tons of compost. Where 2 tons of compost were used, yields were more than double the non-amended plots. Vith the sludge treatment, yields were comparable or slightly higher than where no mendment was added. Rhizoctonia root rot caused severe infections, plant stunting, nd premature death where no compost was applied. Plants growing under the sludge eatment suffered severe root infection. Disease was reduced considerably as compost ates increased from 36 to 72 tons per acre.

Compost is effective because it fosters a more diverse soil environment in which myriad of soil organisms exist. Compost acts as a food source and shelter for the ntagonists that compete with plant pathogens, for those organisms that prey on and arasitize pathogens, and for those beneficials that produce antibiotics. Root rots caused y Pythium and Phytophthora are generally suppressed by the high numbers and iversity of beneficial microbes found in the compost. Such beneficials prevent the ermination of spores and infection of plants growing on the amended soil. To get more eliable results from compost, the compost itself needs to be stable and of consistent uality.

Salinas, California, strawberry grower Tom Jones uses compost on his 125 acres of erries. Eight of those acres have had compost for two years and have just been certified rganic. He uses microbially active, top-quality compost at an eight-ton per acre rate n the organic field and at five-tons per acre on the rest of the fields. Supplemented sh emulsion and other liquid fertilizers are supplied through a drip irrigation system.

Wisconsin fruit and vegetable farmers Richard DeWilde and Linda Halley have rown organic vegetables since 1991. University scientists are doing research on their rm to determine the effect of compost on health and productivity of vegetable crops nd the soil microbial community. DeWilde makes quality compost on-farm from dairy nd goat manure, applied at 10 to 15 tons per acre, realizing a 10% yield increase in ne year. Systemic resistance is also induced in plants in response to compost eatments. Hoitink has now established that composts and compost teas indeed activate isease resistance genes in plants. These disease resistance genes are typically "turned n" by the plant in response to the presence of a pathogen. These genes mobilize nemical defenses against the pathogen invasion, although often too late to avoid the sease. Plants growing in compost, however, have these disease-prevention systems ready running. Induced resistance is somewhat pathogen-specific, but it does allow n additional way to manage certain diseases through common farming practices.

It has become evident that a "one size fits all" approach to composting used in disease anagement will not work. Depending on feed stock, inoculum, and composting ocess, composts have different characteristics affecting disease management potential.

For example, high carbon to nitrogen ratio (C:N) tree bark compost generally works well to suppress Fusarium wilts. With lower C:N ratio composts, Fusarium wilts may become more severe as a result of the excess nitrogen, which favors Fusarium.. Compost from sewage sludge typically has a low C:N ratio.

Some of the beneficial microorganisms that re-inhabit compost from the outside edges after heating has subsided include several bacteria (Bacillus species, Flavobacterium balustinum, and various Pseudomanas species) and several fungi (Streptomyces, Penicillin, Trichoderma, and Gliocladium verens). The moisture content following peak heating of a compost is critical to the range of organisms inhabiting the finished compost. Dry composts with less than 34% moisture are likely to be colonized by fungi and, therefore, are conducive to Pythium diseases. Compost with at least 40 to 50% moisture will be colonized by both bacteria and fungi and will be disease suppressive. Water is typically added during the composting process to avoid a dry condition. Compost pH below 5.0 inhibits bacterial biocontrol agents. Compost made in the open air near trees has a higher diversity of microbes than compost made under a roof or in-vessel.

Three approaches can be used to increase the suppressiveness of compost. First curing the compost for four months or more; second, incorporating the compost in the field soil several months before planting; and third, inoculating the compost with specific biocontrol agents. Two of the more common beneficials used to inoculate compost are strains of Trichoderma and Flavobacterium, added to suppress Rhizoctonia solani. Trichoderma harzianum acts against a broad range of soil-borne fungal crop pathogens, including R. solani, by production of anti-fungal exudates.

The key to disease suppression in compost is the level of decomposition. As the compost matures, it becomes more suppressive. Readily available carbon compounds found in low-quality, immature compost can support Pythium and Rhizoctonia. As these compounds are reduced during the complete composting process, saprophytic growth of these pathogens is dramatically slowed. Beneficials such as Trichoderma hamatum and T. harzianum, unable to suppress Rhizoctonia in immature composts are extremely effective when introduced into mature composts.

For Pythium suppression, there is a direct correlation between general microbial activity, the amount of microbial biomasss, and the degree of suppression. Pythium is a nutrient-dependent pathogen with the ability to colonize fresh plant residue, especially in soil that has been fumigated to kill all soil life. The severity of diseases caused by Pythium and R.solani relates less to the inoculum density than to the amount of saprophytic growth the pathogen achieves before infection. Consequently, soils that are antagonistic to saprophytic growth of Pythium—such as soils amended with fully decomposed compost—will lower disease levels.

Rhizoctonia is a highly competitive fungus that colonizes fresh organic matter. Its ability to colonize decomposed organic matter is decreased or non-existent. There is a direct relationship between a compost's level of decomposition and its suppression of Rhizoctonia—again pointing to the need for high-quality, mature compost. Like immature compost, raw manure is conducive to diseases at first and becomes suppressive after decomposition. In other words, organic amendments supporting high biological activity (i.e., decomposition) are suppressive of plant-root diseases, while raw organic matter will often favor colonization by pathogens.

Determining and Monitoring Compost Quality

It is clear that compost maturity is a key factor in its ability to suppress disease. The challenge involved in achieving and measuring that maturity is the primary reason that compost is not more widely used. Certainly, immature compost can be used in field situations, as long as it is applied well ahead of planting, allowing for eventual stabilization. However, good disease suppression may not develop due to other factors. For example, highly saline compost actually enhances Pythium and Phytophthora diseases unless applied months ahead of planting to allow for leaching.

Dr. Harry Hoitink at Ohio State University has pioneered much of the work associated with disease suppressive composts. He notes that success or failure of any compost treatment for disease control depends on the nature of the raw product from which the compost was prepared, the maturity of the compost, and the composting process used. Failure to assess compost quality may be responsible for some of the failures in using compost for disease suppression. High-quality compost should contain disease-suppressive organisms and mycorrhizal innoculum. Furthermore, high-quality compost should contain very few if any weed seeds.

Several companies offer compost quality testing. Some of these also offer training on how to produce disease-suppressive compost. BBC Laboratories offers a pathogen inhibition assay. Using this assay can determine the ability of your compost sample to directly inhibit specified soil-borne pathogens, including Fusarium, Phytophthora, Pythium, and Rhizoctonia. Each assay costs $75 and tests 12 replicates of your compost compared to 12 replicates of a control where the disease organism is uninhibited. They test for a number of other pathogens in addition to those mentioned above. They can test compost for microbial functional groups such as anaerobes, aerobes, yeasts, molds, actinomycetes, pseudomanads, and nitrogen-fixing bacteria. Their diversity analysis test looks at how many different kinds of organisms exist within each functional group. This information provides insight into how diverse the microbial populations in your compost are. They also test for compost maturity, which determines possible toxicity of immature compost to plants.

Midwest Biosystems offers a grading system for compost quality. Their compos grades range from A to D, with A being disease suppressive. From your submitte sample they will test for sulfates and sulfides, pathogens, nitrogen forms, C:N ratio, see germination, pH, conductivity, redox potential, sodium and moisture levels. This tes costs $30 per sample. An additional test for aerobic plate count and seed germinatio costs $10. Compost grades are assigned based on these tests. For a compost to grad A it must contain 600 to 900 ppm nitrates, no sulfides, meet all lab test guidelines, hav a pH from 7.0 to 8.1, and have a 70 to 100% seed germination rate in pure compos Soil Foodweb, Inc. offers microbial assays including microorganism diversity an biomass. They will comment on the disease suppressiveness of a compost sample base on their performance database for highly productive compost and the plant you ar planning to put the compost on.

DIRECT INOCULATION WITH BENEFICIAL ORGANISMS

There are a number of commercial products containing beneficial, disease-suppressiv organisms. These products are applied in various ways—including seed treatments compost inoculants, soil inoculants, and soil drenches. Among the beneficial organism available are Trichoderma, Flavobacterium, Streptomycetes, Gliocladium spp., Bacillu spp., Pseudomonas spp., and others. Trichoderma and Gliocladium are effective a parasitizing other fungi, but they stay alive only as long as they have other fungi t parasitize. So, these fungi do a good job on the pathogenic fungi that are present wher you inoculate them, but then they run out of food and go to sleep. In soils with low fungal biomass (soils with low organic matter and plenty of tillage) these two beneficial have nothing to feed on. Compost is a great source of both the organisms and the foo they need to do their jobs. A great diversity of bacteria, fungi, protozoa and beneficia nematodes exists in good compost.

REFERENCES

Ingham, Elaine. 1998. "Replacing methyl bromide with compost". *BioCycle*. December. p. 80–82.

Schneider, R.W. (ed.). 1982. *Suppressive Soils and Plant Disease*. The American Phytopathological Society St. Paul, MN. 88 p.

Ingham, Elaine R. 1991. "Interactions among mycorrhizal fungi, rhizosphere organisms, and plants" p. 169–197. In: P. Barbosa, V.A. Krischik, and C.G. Jones (eds.). *Microbial Mediation of Plant-Herbivor Interactions*. John Wiley and Sons, Inc. 530 p.

Maronek, D.M. 1981. "Mycorrhizal fungi and their importance in horticultural corp production". p. 172 213. In: Jaurch, R., *Horticultural Review*. Volume 3.

Matsubara, Yoh-ichi, Haruto Tamura, and Takashi Harada. 1995. "Growth enhancement and verticilliun wilt control by vesicular-arbuscular mycorrhizal fungus inoculation in eggplant". *Journal o Japanese Horticultural Society*. Volume 64, Number 3. p. 555-561.

Newsham, K.K., A.H. Fitter, and A.R. Watkinson. 1995. "Arbuscular mycorrhiza protect an annual gras from root pathogenic fungi in the field". *Journal of Ecology*. Volume 83. p. 991–1000.

2

Soil Ecology

Soil is made up of a multitude of physical, chemical, and biological entities, with many interactions occurring among them. Soil is a variable mixture of broken and weathered minerals and decaying organic matter. Together with the proper amounts of air and water, it supplies, in part, sustenance for plants as well as mechanical support. The diversity and abundance of soil life exceeds that of any other ecosystem. Plant establishment, competitiveness, and growth is governed largely by the ecology below-ground, so understanding this system is an essential component of plant sciences and terrestrial ecology.

An incredible diversity of organisms make up the soil food web. They range in size from the tiniest one-celled bacteria, algae, fungi, and protozoa, to the more complex nematodes and micro-arthropods, to the visible earthworms, insects, small vertebrates, and plants. As these organisms eat, grow, and move through the soil, they make it possible to have clean water, clean air, healthy plants, and moderated water flow.

There are many ways that the soil food web is an integral part of landscape processes. Soil organisms decompose organic compounds, including manure, plant residues, and pesticides, preventing them from entering water and becoming pollutants. They sequester nitrogen and other nutrients that might otherwise enter groundwater, and they fix nitrogen from the atmosphere, making it available to plants. Many organisms enhance soil aggregation and porosity, thus increasing infiltration and reducing surface runoff. Soil organisms prey on crop pests and are food for above-ground animals.

Research interests span many aspects of soil ecology and microbiology, Fundamentally, researchers are interested in understanding the interplay among microorganisms, fauna, and plants, the biogeochemical processes they carry out, and the physical environment in which their activities take place, and applying this knowledge to address environmental problems.

Soil organisms are an integral part of agricultural ecosystems. The presence of a range of soil organisms is essential for the maintenance of healthy productive soils. An excessive reduction in soil biodiversity, especially the loss of species with unique functions, may have catastrophic effects, leading to the long-term degradation of soil and the loss of agricultural productive capacity. As a consequence, more land would be needed for agricultural production to meet demands. The overlooking and depletion of the beneficial functions performed by soil organisms in agricultural ecosystems as a result of inappropriate soil biological management is contributing to increased rates of land degradation, nutrient depletion, fertility decline, water scarcity, and yield reductions. All these factors have a negative impact on the livelihoods of people who depend directly on agriculture for their subsistence.

One of the main gaps in most agricultural management systems is their failure to consider the option of managing soil biological processes and, in particular, using practices that favour the activity of soil macrofauna as a means to maintain and improve soil fertility. Although not readily visible, relatively more attention has been placed in research and development on the functions of soil micro-organisms—both their positive effects on nutrient cycling and uptake, and the negative effects of soil borne pests, including nematodes (microfauna), and pathogens.

LAND

Sufficiency in global food supply is dependent on the intensification of agriculture. As intensification occurs, chemical and mechanical inputs alter, and often substitute, the biological regulation of soil processes. Inadequate soil management is the principal factor behind the worldwide decline in agricultural productivity. Most modern cropping systems require substantial regular inputs of nutrients to replace those removed through harvesting and burning of residues as well as through leaching and erosion. They also require appropriate tillage and weed management practices to reduce risks of soil compaction and erosion and weed competition; and increasingly no- or reduced-tillage practices are being promoted.

In the same way, continued grazing in pastoral environments soon leads to soil degradation through nutrient depletion and compaction unless appropriate legume-grass combinations and grazing rotations are used. In addition, losses and stresses imposed by chemical contamination through use of herbicides and pesticides, as well as chemical imbalances through soil acidification or salinisation may result in impaired soil biological functioning. Faced with such pressures, soils are a threatened resource. The maintenance or amelioration of soil fertility is an essential factor in the development of sustainable and productive agricultural systems in the long-term. This requires the integration of knowledge of biological processes into the design of land-management systems.

Many systems of agricultural management are not sustainable in the longer term because of the pressures they place on the soil. Production levels may frequently be set on the basis of economic goals rather than the capacity of the soil to withstand particular stresses. Conversely greater attention in agricultural systems to managing the soil biological processes, through providing a beneficial environment for soil macrofauna, can restore soil health and improve soil fertility. Factors in soil formation and functioning and their effects at different levels Soil formation depends on five main factors: climate; parent material; topography; and time; as well as living organisms. With the exception of time, the soil-forming factors are considered as interdependent with multiple feedback effects occurring between them. Furthermore, a hierarchy exists with climate playing a dominant role over parent materials and topography.

The soil-forming factors operate wherever soil and soil-forming materials occur, although their relative influences differ between soils and with location on the earth's surface. The soil-forming factors do not influence the soil directly but act through the medium of soil processes. These processes act in potentially different combinations in each environment. Their many combinations and degrees of expression are reflected in the wide diversity of soils found on the surface of the earth.

CLIMATE

The regional climate is the dominant factor affecting the formation of all soils. It interacts with and conditions the effects of the other factors in determining the biota that can survive in particular environments and the seasonality of its activities. Through its control of temperature and moisture regimes, climate determines the phase of the soil water and the intensity of water fluxes. It thereby controls the transport of solid particles and dissolved materials within developing soils, over their surfaces and laterally in the landscape.

PARENT MATERIAL

Parent material is the basic inorganic material from which the soil is formed. Depending on its physical, chemical and mineralogical composition, it will have a strong influence on the composition and texture of the resulting soil. The type of the parent rock affects soil formation, particularly through: the amount of clay that can be potentially formed by in situ weathering—this varies with the nature of the parent rock and intensity of water fluxes; the amount of alkali, notably sodium (Na) and potassium (K), of alkaline-earth metals, largely calcium (Ca) and magnesium (Mg), and of iron (Fe) that can be released by weathering; the ease of release of the above minerals; the permeability of the parent rock.

TIME

Soils undergo extended and complex series of reactions and processes during formation from their parent materials. The net effect of these eventually leads to the differentiation of fully-developed profiles. This occurs at widely variable rates depending largely on parent materials and environment. In a general way, soils can be divided into those forming over short cycles and long cycles. The short-cycle soils develop over periods ranging from less than 10^3 to 10^4 years. The long-cycle soils require periods of from 10^5 to 10^6 years for development. In any case, the long time necessary for soil formation means that it can be considered a non-renewable resource.

ORGANISMS

The organisms (or biota) are a major factor in soil formation and their effects determine many differences between soils. The various soil organisms affect certain soil processes in different ways. Soil macrofauna play a particularly important role in soil aggregation and porosity as a consequence of their burrowing and mixing activities. This in turn affects the environment (aeration, soil moisture, etc.) for other soil organisms.

Soil is a still, porous, semi-aquatic medium within which temperature and moisture conditions are highly buffered. Soils were among the first terrestrial environments to be colonised because they possess environmental conditions that are intermediate between aquatic and aerial media. Soil is a large reservoir of biodiversity, often little known. Soil communities are among the most species-rich compartments of terrestrial ecosystems. It is believed that there are twice as many species of organisms living in soil than there are in tropical rainforest canopies. Soil organisms carry out a range of processes that are important for soil health and fertility in soils of both natural ecosystems and agricultural systems. They perform and regulate a major proportion of the organic matter transformations and of the carbon and nutrient fluxes in terrestrial ecosystems. The diversity of life in soil, known as soil biodiversity, is an important but poorly understood component of terrestrial ecosystems. Soil biodiversity is comprised of the organisms that spend all or a portion of their life cycles within the soil or on its immediate surface.

The easiest and most widely used system for classifying soil organisms is to group them by size into three main groups: macrobiota, mesobiota and microbiota.

Microbiota comprises microorganisms and microfauna. Microorganisms are the smallest of the soil animals ranging from 20 to 200 μm in length (< 0.1 mm in diameter). They are extremely abundant and diverse. They include: algae, bacteria, cyanobacteria, fungi, yeasts, myxomycetes and actinomycetes that are able to decompose almost any existing natural material. Microorganisms transform organic matter into plant nutrients that are assimilated by plants.

Microfauna includes small Collembola and mites, nematodes, and protozoa, among others, that generally live in the soil-water film and feed on microflora, plant roots, other microfauna and sometimes larger organisms (e.g. entomopathogenic nematodes, that feed on insects and other larger invêrtebrates). Microfauna form the link between the primary decomposers (i.e. microorganisms) and the larger fauna in the detritus food-web in the soil. They are also important to the release of nutrients immobilised by soil microorganisms. The main soil animals in this group are protozoa.

The mesofauna is the next largest group and the animals range in size from 200 μm to 10 mm. in length (0.1–2 mm in diameter). These include mainly microarthropods, such as pseudoscorpions, protura, diplura, springtails, mites, small myriapods (*Pauropoda* and *Symphyla*) and the worm-like enchytraeids. Mesofauna organisms have limited burrowing ability and generally live within soil pores, feeding on organic materials, microflora, microfauna and other invertebrates.

This module focuses specifically on soil macrofauna communities. The macrofauna contains the largest soil invertebrates. Although the term soil macrofauna is not well defined, for the purposes of this training material it will be defined at a higher taxonomic level using the broad criterion:

> "A soil macrofauna taxon (group) is an invertebrate group found within terrestrial soil samples which has more than 90 percent of its specimens (individuals) in such samples visible to the naked eye".

Soil macrofauna groups include organisms like earthworms, millipedes, centipedes, ants, Coleoptera (adults and larvae), Isopoda, spiders, slugs, snails, termites, Dermaptera, Lepidoptera larvae and Diptera larvae.

In terms of their abundance and their soil forming roles, earthworms, termites and ants are the most important macrofauna components of soils. Indeed, the importance of their activities has caused them to be called "ecosystem engineers".

They burrow and are important in mixing the soil known as bioturbation. Macroarthropods and Mollusca are constant inhabitants of litter and, to a lesser extent, of soils, but they have generally more specific ecological roles. Thus, most live in the litter or in the upper few centimetres of soil; saprophagous arthropods play a major role in the breakdown of surface litter.

Most soil animals occur in the top 30 cm of soil, although some also occur at depth. Soil animals may move to lower soil layers when conditions at the surface are harsh. Most soil animals occur in the surface layer because this layer contains the most food (C and nutrients) in the form of organic matter and other organisms. In both natural and agricultural systems, soil organisms perform vital functions in the soil. The interactions among organisms enhance many of these functions, which are often controlled by the enormous amount of organisms in soils. These functions range from physical effects, such as the regulation of soil structure and edaphic (in soil) water regimes, to chemical and biological processes such as degradation of pollutants, decomposition, nutrient cycling, greenhouse gas emission, carbon sequestration, plant protection and growth enhancement or suppression. To reduce the huge complexity of organisms that live in the soil, a division of soil organisms into functional groups has been proposed.

Role of Macrofauna

A functional group consists of a group of organisms that have the same function and similar impact on soil. There is no single classification system because the criteria used to classify soil organisms and the degree of subdivision applied are a function of the questions being addressed. A simple classification is proposed here to assist users of this manual in better understanding the main types of organisms and subcategories according to their main visual characteristics and functions.

The functions that soil organisms in the macrofauna category carry out depend largely on the efficiency of their digestive systems (which themselves depend on their interactions with soil microorganisms, e.g. bacteria) and on the occurrence and abundance of the biological structures that they produce in the soil. Using these two criteria, three large functional groups of invertebrates can be distinguished: micropredators, litter transformers, and ecosystem engineers. The micropredator group contains the smallest invertebrates, protozoa and nematodes. They do not produce organo-mineral structures, and their principal effect is to stimulate the mineralisation of soil organic matter (SOM).

In the litter-transformer group, mesofauna and some macrofauna organisms are involved in litter decomposition. When these invertebrates re-ingest their excretions, which serve as incubators for bacteria, they assimilate metabolites liberated by microbial actions. The "ecological engineers" or "ecosystem engineers" are those organisms that produce physical structures through which they can modify the availability or accessibility of a resource for other organisms. Among the innumerable life forms that inhabit soils, only a small number of macroinvertebrates are distinguished by their capacity to excavate soil and produce a wide variety of organo-mineral structures, such as excretions, nests, mounds, macropores, galleries and caverns. Their structures have been described as "biogenic structures". Their activities and biogenic structures can modify the abundance or structure of their communities. The functional role of these structures is thought to be important because they represent sites where certain pedological processes occur: stimulation of microbial activity; formation of soil structure; SOM dynamics; and exchange of water and gases.

Decomposition and nutrient cycles, hence organic matter dynamics. Soil structure: The activities of certain organisms affect soil structure, porosity and aggregation—especially the "soil engineers" such as worms and termites—through mixing soil horizons and organic matter and increasing porosity. Carbon sequestration and gas exchange: The activities of certain organisms determine the carbon cycle—the rates of carbon sequestration and greenhouse gases (GHGs). Soil hydrological processes, in relation to effects on soil structure and porosity.

Control of pests and diseases: Certain soil organisms can be detrimental to plant growth, e.g. the buildup of nematodes under certain cropping practices. However, they can also protect crops from pest and disease outbreaks through biological control and reduced susceptibility. Soil detoxification: Soil organisms can also be used to reduce or eliminate environmental hazards resulting from accumulations of toxic chemicals or other hazardous wastes. This action is known as bioremediation. Plant production: Plant roots, through their interactions with other soil components and symbiotic relationships, especially Rhizobium bacteria and Mycorrhiza, play a key role in the uptake of nutrients and water, and contribute to the maintenance of soil porosity and organic matter content, through their growth and biomass.

The loss of biodiversity is a dramatic manifestation of the poor management of natural resources. The biological impoverishment caused by inappropriate management could affect the continuous dynamic functioning of ecosystems. It is important to preserve biodiversity in order to maintain the integrity of the processes that sustain the ecosystem services, such as primary productivity, nutrient cycling and consumption of oxygen. Biodiversity is also important to maintaining resilience, i.e. the soil capacity to recuperate its initial situation after a natural or human-induced perturbation. Thus, a system that is functioning properly is one that will persist despite natural environmental fluctuations. There are plausible arguments for an increase in stability increase in diversity such as greater numbers of functionally interchangeable species or species groups, each susceptible to slightly different perturbations, or greater segregation of species into compartments that interact little if at all. In this regard, the removal of any species may increase the susceptibility of the system to the perturbation which may be due to a natural event, such as climatic variability, or human induced, for example toxicity of an agrochemical or effects of severe compaction due to repetitive tillage.

SOIL ECOSYSTEM

Soils are the part of the earth's surface, which forms a narrow interface between the atmosphere and the lithosphere. Soils are made up of water, gases and mineral matter together with a diverse range of organisms and materials of biological origin. Organic materials in and on the soil are broken down and transformed—mainly by soil organisms– into nutrient elements, which are, in turn taken up by plants and micro-organisms. Soil organisms are the main mediators of soil functioning at different scales. These functions can be pictured as having a hierarchical relationship. Figure 1 illustrates the hierarchical organisation of the determinants of soil processes: climate, soil characteristics—especially the abundance and types of clays and nutrients—and the quality of the organic materials input. The series of factors affecting soil functioning are determined by both spatial and temporal scales.

Of major importance in ecosystem and soil development and maintenance are the so-called "ecosystem engineers"—as these species control, either directly or indirectly, the availability of resources to other species. These organisms physically modify, maintain and create new habitats for other organisms. One effect of such organisms is to create higher habitat diversity, which may in turn increase species diversity. An example of physical ecological engineers is plant roots that create large voids (spaces) in the soil through root decay. Other ecosystem engineers are the termites and earthworms that play a major role in moving, mixing and aerating the soil through their burrowing. Other organisms including higher plants and animals also play substantial roles in this respect.

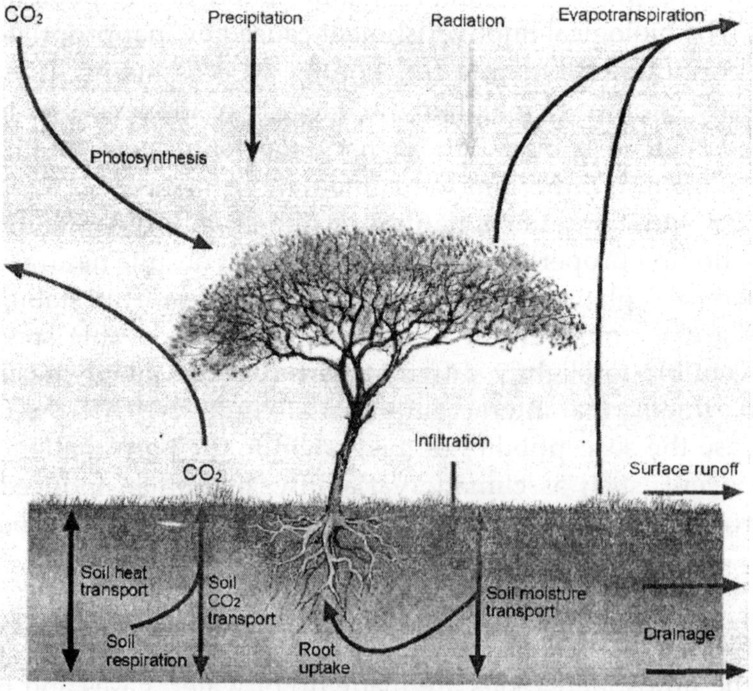

Figure 1. Determinants of soil processes

The functioning of the soil system is also determined by: the decomposition rates of dead organic materials, and the balance between mineralisation, which releases nutrients available to plants and microorganisms, and humification, which forms reserves of soil organic matter (SOM) and colloidal organic compounds; the degree of synchronisation of nutrient release with plant demand; the soil physical structure, which determines the rates and patterns of gas exchange, soil water movement into and through the soil, and erosion rates. The texture of the soil (% of sand, silt and clay) which influences the activity of soil organisms and hence the soil biological functioning. Texture is an important characteristic of soil because it influences many aspects of soil fertility, especially the amount of water held by the soil, its capacity to retain plant nutrients, and the ability of roots to develop and grow through the soil. Soils with a high percentage of clay are said to be "heavy" soils and have a capacity to retain water due to the small pore spaces and high surface tension forces. Soils with a high percentage of sand are considered "light" soils, and tend to hold very little water. Water infiltrates rapidly into sandy soils and is readily drained through the large pores spaces, unless they also contain a lot of organic matter.

SOIL HEALTH

Soil is usually viewed simply as a medium for growing plants. However, in addition to providing a mechanical support for plants, soil enables the storage of water and organic matter, releases elements of biological and pedological importance, and is the place where soil organisms live.

Soil health refers to the capacity of soil to function as a vital living system, a dynamic system. It involves the idea of soil as a living dynamic organism that functions in a holistic way depending upon its condition or state. Soil health depends on the combined effects of three major interacting components. These are the chemical, physical and biological characteristics of the soil. Soil health is enhanced by management and land-use decisions that consider the multiple functions of soil. It is impaired by decisions that focus only on single functions and short-term solutions, such as increasing but not sustaining crop productivity.

Maintaining and improving the capacity of soils to function is essential to human survival, and healthy soil is an essential element within this process. The ecological attributes of the soil are important since they have implications beyond the quality of the soil or its health, the capacity to produce a particular crop. They are associated with the soil biota, its diversity, its food-web structure, its activity and the range of functions it performs in the system. The soil biota is a vital force that serves to maintain the health of the soils. Soil biodiversity per se is not a property of soil that is critical for the production of a given crop, but it may be very important for the continued capacity of the soil to produce that crop.

Soil organisms exert a major control over many soil processes through their effects on: the decomposition of dead organic materials; nutrient cycling; the modification and transport of soil materials; and the formation and maintenance of soil structure. Although sometimes not easily recognised, the biological activity of soils is largely concentrated in the topsoil, from a few centimetres to 30 cm. The living component of the SOM consists of: plant roots—5–15 percent, soil organisms—85–95 percent; and their components: macrofauna and mesofauna—15–30 percent; microorganisms—60–80 percent.

Much of the biological activity is associated with processes that regulate nutrient cycling (mineralisation, denitrification, nitrogen fixation, etc.) and the decomposition of organic residues. Soil macrofauna has an important role in these processes as dead organic matter is first consumed by macrofauna that digest it, partially cutting it into small fragments. These small fragments and partially degraded organic residues are then available to mesofauna and microorganisms. Through its burrowing activity, soil macrofauna buries organic matter deep into the soil and stimulates the activity of microorganisms. Thus, soil organisms participate in a range of processes essential to the functioning of ecosystems. They also constitute an important resource for the sustainable management of agricultural ecosystems and their durability. For example, they play an important role in the dynamics of SOM and nutrient cycling, the purification of water, the detoxification of agrochemicals, and the modification of soil structure. Macroorganisms and microorganisms exert a strong influence on soil properties through the role they play in organic matter decomposition, nutrient redistribution and cycling, and the transformation of such nutrients into forms readily available for plant nutrition. Soil biota and their biological activities are important for soil health and fertility in agricultural soils. The activities of soil organisms interact in a complex food-web. The soil food-web is a way to relate soil organisms to one another on the basis of what they eat. Some of these organisms feed on living plants (herbivores) and animals (predators), some on dead plant debris (detritivores), some on fungi or bacteria, and others live off, but without consuming, their hosts (parasites).

Plants, mosses and some algae are autotrophs. They play the role of primary producers by using solar energy, water and C from atmospheric carbon dioxide (CO_2) to make organic compounds and living tissues. Other autotrophs obtain energy from the breakdown of soil minerals—oxidation of nitrogen (N), sulphur (S), iron (Fe) and carbon (c) and from carbonate minerals. Soil fauna and most fungi, bacteria and actinomycetes are heterotrophs, they rely on organic materials directly (primary consumers) and through intermediaries (secondary or tertiary consumers) for C and energy needs. All terrestrial ecosystems, including agricultural production systems, consist of a producer subsystem (the crop), and a decomposer subsystem, and both components depend upon each other.

A biologically healthy soil harbours a multitude of different organisms—microorganisms, such as bacteria, fungi, amoebae and paramecia, as well as larger organisms such as nematodes, springtails, insect larvae, ants, termites, earthworms and ground beetles. Most are helpful to plants, enhancing the availability of nutrients and producing chemicals that stimulate plant growth. A healthy soil produces healthy crops with minimal amounts of external inputs and few to no adverse ecological effects. It has favourable biological, physical and chemical properties.

Among the functions regulated by soil organisms, the most important are: decomposition: breaking down litter, creating humus and cycling nutrients; converting atmospheric N into organic forms and reconverting organic N into gaseous N; synthesizing enzymes, vitamins, hormones and other important substances for plant growth; modifying soil structure, thus affecting porosity, water fluxes and organic matter distribution and promoting deeper root growth; eating and/or decomposing weed seeds; suppressing and/or feeding on soil-borne plant pathogens and plant-parasitic nematodes.

Healthy soils are also essential to plant defences. Unhealthy soils hinder crops' abilities to use their natural defences and leave them vulnerable to potential pests. In contrast, healthy soils arm plants chemically with defence-boosting nutrients and are physically conducive to optimal root development and water use. Healthy soils can also expose weed seeds to more predators and decomposers, and their slower release of N in spring can delay small-seeded weeds (which often need a flush of N to germinate and begin rapid growth), thereby giving larger-seeded crops a head start.

Soil health can be improved by: diversifying crop rotations including legumes and perennial forages; keeping soils covered year-round with living vegetation and/or crop residue; adding plenty of organic matter from animal manures, crop residues and other sources to restore that removed through harvest or lost through burning and breakdown; reducing tillage intensity and protecting soils from erosion and compaction; using best management techniques to supply balanced nutrients to plants without polluting water.

Consequently, a main objective of every farmer should be to support high levels of potentially beneficial soil organisms and low levels of potentially harmful ones. A soil rich in fresh organic residues –sometimes called particulate or light-fraction organic matter—can feed huge numbers of organisms and foster abundant and diverse biological activity.

REFERENCES

Bollen, Walter B. 1959. *Microorganisms and Soil Fertility*. Oregon State College. Oregon State Monographs, Studies in Bacteriology, No. 1. 22 p.

Cook, R.J., and K.F. Baker. 1983. *The Nature and Practice of Biological Control of Plant Pathogens.* APS Press, St. Paul, MN.

Engelhard A.W. (ed.). *Soilborne Plant Pathogens: Management of Diseases with Macro- and Microelements.* American Phytopathological Society. 217 p.

Hudson, Berman D. 1994. "Soil organic matter and available water capacity". *Journal of Soil and Water Conservation.* March–April. p. 189-194.

Jackson, William R. 1993. *Organic Soil Conditioning.* Jackson Research Center, Evergreen, CO. 957 p.

Johnston, S.A., and P.J. Nitzche. *Rotation periods suggested to help control vegetable diseases.* New Jersey Extension Service. 1 p.

Logsdon, Gene. 1995. *Using compost for plant disease control. Farm Scale Composting.* JG Press, Inc. Emmaus, PA.

Ozores-Hampton, Monica, H. Bryan, and R. McMillian Jr. 1994. "Suppressing disease in field crops". *BioCycle.* July. p. 60–61.

Schneider, R.W. (ed.). 1982. *Suppressive Soils and Plant Disease.* The American Phytopathological Society. St. Paul, MN. 88 p.

Sances, Frank V., and Elaine R. Ingham. 1997. "Conventional organic alternatives to methyl bromide on California strawberries". *Compost Science and Utilization.* Spring. p. 23–37.

3

Role of Soil Macrofauna

Soil macrofauna consists of a large number of different organisms that live on the soil surface, in the soil spaces (pores) and in the soil area near roots. Their way of living, their feedings habits, their movements into the soil, their excretions and their death have direct and indirect impacts on their habitat. The biological activities of soil macrofauna regulate soil processes and soil fertility to a significant extent. The effects of soil macrofauna on soil can be divided into three classes: physical, chemical and biological effects. These effects are determined by the functional group involved in the process.

PHYSICAL ROLE OF SOIL MACROFAUNA

Five main physical effects of soil macrofauna can be highlighted:

- macromixing,
- micromixing,
- gallery construction,
- fragmentation,
- aggregate formation.

Macromixing

Ants, termites, earthworms and ground beetles can move an important quantity of soil, bringing back to the surface mineral matters from deeper horizons and burying the organic matter from the surface horizons, from litter and from excrements. For example, a large nest of Atta ants comprises several million individuals. It forms a cavity in soil with numerous chambers. The excavated earth is deposited on the soil surface surrounding the nest. The removal of fine material in depth sometimes creates porous zones under the nest where water can be accumulated temporarily. The macromixing activity of earthworms is of major importance to soils. It can be measured by the quantity of casts found on the soil surface. Earthworms can produce 40–250 tonnes of casts per hectare per year. Some can produce up to 2 500 tonnes of casts per hectare per year.

Some beetles (especially those of the subfamily Scarabeidae) are coprophagous – they are very efficient at incorporating and removing excrements that are on the soil surface. For example, just a couple of Heliocopris dilloni, a large African species, can bury a piece of dung in one night.

Micromixing

Other groups of soil macrofauna influence soil structure in a less spectacular way, but the micromixing that they realise is as important as macromixing. These organisms, mainly represented by Diptera larvae, have a more limited capacity to dig the soil. They stay on the soil surface where they realise a fundamental task for the incorporation of organic matter to soil. However, they can be carried into soil by leaching to a depth of up to 60 cm.

Gallery Construction

Gallery (burrow) formation is very important for soil aeration and water flux. For example, earthworms and termites develop networks of galleries that improve large spaces in the soil macro-porosity by 20–100 percent. Earthworms can burrow an estimated 400–500 m of galleries per square metre in grasslands. These galleries are denser in the top 40 cm and can represent up to 3 percent of the total soil volume. In these conditions, the waterholding capacity of soil can increase by 80 percent and water flux can be from four to ten times faster.

Earthworm activity is very important in agricultural soils with a high degree of compaction and a ploughing pan that prevents water flux. This situation decreases water infiltration and increases surface runoff and erosion. Earthworms pierce the ploughing pan, so improving water infiltration and offering new paths for root penetration. Termite excavation activity has a similar effect on soils, and in some cases can reduce the compaction of surface layers. Where organic matter is present in the soil, the bioturbating and decomposing activities of termites can reduce soil compaction, increase its porosity and improve its water infiltration and retention capabilities. Such conditions encourage root penetration, vegetative diversity and the restoration of primary productivity.

Thus, galleries make up a draining system that collects rainwater and facilitates its flow. Water drags small material into these tunnels, which become the preferential paths for soil penetration for roots and leached clays. Galleries are also the soil penetration paths for other surface invertebrates with more limited burrowing capacities, e.g. very small earthworms, slugs, insect larvae, and mesofauna. The fragmentation of dead wood (lignin material), carcass and litter is one of the most important activities of soil fauna. It has a major effect on organic matter evolution in soil, conditioning the activity of bacteria, fungi and microfauna populations. Fragmentation is performed by phytosaprophagous animals (i.e. animals feeding on decayed plant material and dead animals).

Aggregate Formation

After litter has been fragmented, it is easier for organic matter to be broken down into the stable form known as "humus", and then to form soil aggregates – the clumping together of soil particles forming a crumbly healthy structure. Earthworms, termites, millipedes, centipedes and woodlice ingest soil particles with their food and contribute to aggregate formation by mixing organic and mineral matter in their gut.

CHEMICAL EFFECTS

The most important chemical effect of macrofauna on soil is the modification of food quality through its passage in the gut and particularly the mineralisation of organic matter and the release of nutrients. Soil macrofauna also influences soil chemical composition through the deposition of excrement. The main indirect chemical effect is the mineralisation of N, P and S through the activation of microflora. Soil microorganisms represent an important proportion of the soil living component (60–80 percent). However, in order to be active, they need to be in contact with SOM (which they feed on). Because of their inability to move in soil to search for food, microorganisms are only active during short periods of time (the time necessary to consume the organic matter around them) and in a limited number of microsites (where temperature and moisture conditions are suitable for their activity). The rest of the time, soil microorganisms are "in dormancy" and they are able to survive "hard times" in this way.

The contrast between the potential of soil microorganisms for an extremely fast turnover of organic matter and the field reality has been called the "Sleeping Beauty Paradox". Macrofauna that has the ability to move the soil and change environmental conditions at the scale of microorganisms can interrupt this dormancy, providing assimilable substrates (root exudates, earthworm mucus and other materials) that initiate the metabolic capabilities of microorganisms. Hence, they appear to be major regulators of microbial activities. Interactions between microorganisms (with a high capacity to digest almost all organic substrates) and macrofauna (with the potential for mechanical activities) are the basis of the biological systems of regulation that determine soil function.

BIOLOGICAL EFFECTS

In a natural soil, a complex and dynamic balance exists between the different groups of organisms with different feeding habits. Predation and competition are the main factors controlling this equilibrium. Predation has an important role because it establishes a balance between the number of individuals and the quantity of available resources. Competition is another way to maintain soil fauna populations in balance with soil resources.

Another biological effect of soil macrofauna is the disappearance of dead animal material. This work is realised by necrophagous (which feed on dead and/or decaying animals) and coprophagous organisms (feeding on dung or excrement) such as Diptera larvae, Coleoptera and Lepidoptera larvae and adults. They clean the soil surface and incorporate organic matter into soil. In addition, soil macrofauna disseminates bacteria and spores through excrement dispersion in soils or by on-body transport. Earthworms determine the vertical repartition depth in soil...

Thus, through their different activities and effects on soil, soil organisms create and accumulate structures that give soils specific architectures. Networks of galleries, the accumulation and spatial array of biogenic aggregates and surface deposits are among the conspicuous features that can be observed in the field. The nature and array of these structures depend on the organisms that have produced them. The physical and chemical parameters of the soil that was used to make the biostructures are also important as they determine the resistance and persistence of these structures. The sum of structures deposited over time by these organisms have specific textural, structural and architectural properties that influence the physical-chemical properties of soil and the smaller fauna and microflora that live in this environment.

These sets of structures (pores, aggregates, etc.) that have been accumulated by soil organisms can be colonised by rather specific communities of microorganisms, other invertebrates and possibly roots. The environment in these structures can be very different from that of soil. For example, the availability of C, mineral N and P that can be assimilated may be enhanced significantly in fresh earthworm casts or fresh termite pellets when compared with the ingested soil. As a consequence, other groups of organisms such as litter transformers, micropredators, microorganisms and fine roots may become established in these domains.

The importance of the functions performed in soils by macrofauna and the physical, chemical and biological changes induced in a soil environment as a consequence of its activity make it a vital part of all ecosystems, including agro-ecosystems.

Soil macrofauna is involved in:

- — degrading organic matter and mineralising nutrients; controlling pathogen populations;
- — improving and maintaining
- — soil structure;
- — mixing organic matter through the soil.

The reduction of aboveground biodiversity is normally associated with the alteration of several environmental parameters including the carbon supply to the soil, which provides the basis for a more or less diversified soil population. The reduction in diversity of soil organisms causes a dysfunctioning of the ecosystem. This dysfunctioning may result in disequilibrium between beneficial and harmful organisms, which can lead to harmful organisms to become dominant.

In other cases, as result of a reduction in macrofauna, the activities of a beneficial organism can become detrimental. Barros et al. and Blanchart et al. showed the results of dysfunctioning caused by the disequilibrium between compacting and uncompacting macroinvertebrates The activity of the earthworms in the pastures of central Amazonia can be positive where the other groups of the macrofauna remain present. However, as soon as the compacting earthworm *P. corethrurus* becomes largely dominant and the decompacting fauna groups disappear, a compacting activity begins, resulting in the formation of a massive surface layer. Improvements in soil management to avoid a reduction in soil macrofauna diversity and to guarantee suitable soil conditions for these organisms to develop are the only way to ensure long-term soil functioning and to protect cultures from pests.

EFFECT OF LAND-USE AND MANAGEMENT PRACTICES

Macrofauna Composition

Land-use management has an important impact on soil and its functional role in maintaining ecosystem processes. It generally results in dramatic and rapid changes in vegetation that are likely to affect soil invertebrate communities significantly.

The biodiversity of animals and plants generally declines as an inverse function of the intensity with which crops are cultivated using mechanised methods and agrochemicals, that is agricultural intensification. The intensity with which soils are cultivated also depletes soil-organism communities as a consequence of the toxic effects of agrochemicals, the physical disruption of their habitats, and the reduction in litter availability and hence the SOM resource base. Thus, management practices have important consequences on the composition and abundance of soil macrofauna communities.

Lavelle et al. identified the major trends in tropical soil macrofauna composition as follows.

Termites and/or earthworms tend to be dominant in most cases. Termites have adapted to a wide range of semi-arid systems where earthworms are not found. Earthworms are best represented in grasslands in tropical humid areas. They are sensitive to the nutrient status and organic content of soil. Litter arthropods are predominant where sufficient litter is available. They are represented mainly by millipedes or beetles.Lands with natural vegetation like forests have a highly diversified vegetation. The quality and quantity of leaf litter and the important root production create a particular soil environment that favours the development of a rather diverse and abundant soil macrofauna.

When vegetation is cleared and the soil used for cropping, the communities of soil macrofauna are highly depleted as a consequence of an important modification of soil microclimate, the reduction in the quantity and quality of organic matter, the physical perturbation induced by tillage, and the effects of fertilisers and non-targeted pesticides. Earthworms and litter communities soon disappear and are not replaced. Other macrofauna groups such as termites tend to be more persistent. Soils under pasture are more favourable for earthworm development as there is an improvement in leaf-litter quality, a great quantity of manure brought to the soil, and a more limited physical soil perturbation unless seriously overgrazed. Other soil management practices, such as palm tree plantations with herbaceous legume cover or cocoa with a layer at the soil surface and high trees, usually have diverse soil biotic communities.

Thus, land management has a dramatic effect on soil macrofauna communities. Primary forests have rather diverse and abundant fauna with density and biomass two to three times higher than that in managed systems. Epigeic and litter fauna is well represented and the biological activity is mainly concentrated in the top 20 cm of soil.

Cropping results in a dramatic decrease in taxonomic richness, density and biomass. Termites are the major component. Earthworms are mainly represented by endogeic species and their distribution in the soil profile is relatively deep (20–30 cm).

Pastures can have very different communities with an overwhelming dominance of populations of the peregrine (wandering) earthworm Pontoscolex corethrurus. This is an endogeic earthworm frequently present in disturbed areas and it may build up very large populations in pastures. Other groups of importance are termites and ants. Litter-dwelling populations represent only 4.3–10.5 percent of individuals, most of them concentrated in the upper 10 cm. Short fallows present low taxonomic richness but increased densities and a slight evolution towards the original forest situation. The association of palm trees with a leguminous cover crop appears to be the most conservative system with both elements of the primary forest and introduced ones. This kind of system can present a large earthworm community with forest and pasture species and high densities of termites and epigeic litter detritivores, which live and feed on the soil surface.

Positive Impacts on Soil Macrofauna

Some agricultural management practices have positive impacts on soils, increasing SOM levels and improving soil functioning and plant productivity. For example, conservation agriculture can increase crop production while reducing erosion and reversing soil fertility decline. SOM and biological activity in the rootzone are stimulated by continual additions of fresh organic matter (crop residues, cover crops and manure).

Indirect Management Practices

Where indirect management practices are used, interventions are a means of managing soil biotic processes by manipulating the factors that control biotic activity (habitat structure, microclimate, nutrients and energy resources) rather than the organisms themselves. Examples of indirect interventions include most agricultural practices, such as application of organic materials to soil, tillage, fertilisation, irrigation, green manuring and liming as well as cropping system design and management.

In conventional agriculture, soil tillage is considered one of the most important operations for creating favourable soil structure, preparing the seedbed and controlling weeds. However, mechanical implements destroy the soil structure by reducing the aggregate size, and conventional tillage methods are a major cause of soil loss and desertification. A study in France showed that, in fields under reduced or no-tillage and with permanent soil cover, soil organisms can increase by up to four times and the diversity may almost double. In this kind of management system, the biomass produced is kept on the soil surface and serves as a physical protection of the soil and provides food for the soil fauna. In general, the vegetal cover on the soil surface creates a more humid environment, which is conducive to the activity of soil organisms. The larger number of earthworms, termites, ants and millipedes combined with a high density of plant roots results in more large pores, which in turn increase water infiltration.

Pest management can also benefit from conservation practices that enhance biological activity and diversity, and hence competitors and predators as well as alternative sources of food. For example, most nematode species (especially pathogens) can be reduced significantly by the application of organic matter, which stimulates the action of several species of fungi that attack nematodes and their eggs. An agricultural technique used in a number of tropical countries in Africa, Asia and South America to ameliorate soil conditions for crops is "ecobuage". This is a complex agricultural system that entails incinerating herbaceous vegetation piled up in mounds and buried under a layer of soil taken from the surroundings. It is a traditional system that is more evolved than the slash-and-burn technique often used in intertropical zones.

A study conducted in the region of Bouenza (Congo) showed that the use of "Maalas" (ecobuage) improved soil macrofauna communities, particularly earthworm communities. The technique supplied the soil with mineral nutrients through slashes, and increased soil pH. The improvement in soil macroinvertebrate communities improved soil structural stability, creating good conditions for plant root development. Thus, this system enabled the cultivation of plants that are demanding in terms of nutrient supply. In addition, the earthworm activity improved soil porosity (macroporosity became more important), allowing plant roots to go deeper into the soil. Improved soil structure enabled the cropping of plants with tubercules that need an aerated soil for their development. Experiments at Carimagua (Colombia) suggest that the spatial and temporal array of parcels allocated to different crops may favour the conservation of locally high earthworm-population density and diversity. These spots may serve as reservoirs and refuges for the colonisation of depopulated areas.

Direct management practices

These practices involve intervening in the production system in an attempt to alter the abundance or activity of specific groups of organisms. Examples of direct interventions include:

(i) inoculation of seeds or roots with rhizobia, mycorrhizae, fungi and rhizobacteria for enhanced soil fertility; and

(ii) inoculation of soil or the environment with biocontrol agents, antagonists and beneficial fauna (e.g. earthworms).

Mycorrhizal microorganisms colonise the roots of many plants through a symbiotic relationship. This relationship helps the plants to be more efficient at obtaining available nutrients, such as P and N, and water, both elements vital to plant survival. Mycorrhizae increase the surface area associated with the plant root, which allows the plant to reach nutrients and water that might not be otherwise available. The application of N-fixing rhizobia for the production of common beans in the Parana region in Brazil increased yields by almost 50 percent. This kind of intervention might be a way of reversing the poor yields and nitrogen depletion that plague tropical areas. In southern India, the long-term exploitation of soil under tea gardens has led to impoverishment of soil fertility and stabilisation of yields despite increasing applications of external inputs such as fertilisers and pesticides. The application of high-quality organic matter and earthworms was very effective at increasing tea yields (more than by the application of fertilisers alone) owing to their favourable effects on physical and biological soil properties. Yields increased by 79.5–276 percent.

The loss of abundance and diversity of communities under annual crops results invariably in a loss of certain important soil functions.

The results obtained in these studies suggest various options for conserving and stimulating the activities of soil macrofauna. For example, the negative effects of annual crops could be reduced by decreasing the intensity and frequency of perturbations such as tillage and the use of pesticides, and by increasing the quantity and quality of the energy resources used by the macroinvertebrates, e.g. the use of legume cover crops and the maintenance of crop residues. Integrated systems of short phases of crops with longer periods of pastures (3–5 years) are also an option for maintaining macroinvertebrate populations as well as bringing other benefits for soil physical and chemical parameters.

Organic manuring helps to enrich or favour the multiplication of many soil fauna and microorganisms including those antagonistic to soil pests. In recent years, the application of greater quantities of synthetic fertilisers has been very common in contrast to the negligible use of organic manures. The eggs, larvae and pupae of soil insects are liable to be affected either by the soil-inhabiting pathogens or their toxins. However, the absence of organic manures in the soil enables the above pests to thrive owing to the depletion of the natural biotic restricting factors. The increasing use of pesticides has also upset the balance of life in soil when they have been applied directly in the soil and by drip from the foliage. Their subsequent incorporation into soil can also reduce the natural enemies of soil insects.

Certain practices such us improved pasture can result in increased populations of soil macrofauna. The similarity of the original ecosystem and the derived agro-ecosystem tends to be a major determinant of native species' survival, adaptation, resilience and stability within the boundaries of ecosystem management. The spatial arrangement of pastures alongside cropped plots can accelerate the recovery of macrofauna populations in the cropped plots. Beneficial species, which can be more rapidly established, can also help reverse some of the degrading effects of cropping on soil structure, thereby avoiding the need to solve soil degradation problems with expensive, machinery-intensive strategies. Thus, earthworms become a resource that can be harnessed to improve ecosystem health.

Negative Impacts on Soil Macrofauna

In Corrientes Province (northern Argentina), the use of heavy machinery to prepare soil for rice culture has caused soil compaction. This type of soil management has led to important changes in soil macrofauna abundance and diversity. After several years of culture, fields are left fallow. During this period, the soils are recolonised by soil fauna. However, the soil compaction makes this process difficult and the composition of the soil macrofauna is dominated by the ant species Camponotus punctulatus. These ants usually build their nests in soil, but soil compaction modifies their behaviour and they construct epigeic mounds. As a consequence, the soil becomes covered by ant mounds 1–2 m high, which can reach densities of more than 2 000 mounds/ha in 2–3 years. This action completely prevents the possibility of further agricultural uses of the land, without extensive and expensive measures to destroy the ant nests and the ant populations.

ECOLOGY OF SOIL MACROFAUNA

The main groups of soil macrofauna in terms of their abundance and the importance of their activities in soil are: earthworms, termites, ants, Myriapoda, Diptera and Coleoptera.

Earthworms

Earthworms are terrestrial worms that burrow into and help to aerate soil. They often surface when the ground is cool or wet. They have a segmented body. The body may be divided into two parts: an anterior part with segments containing cephalic ganglions, reproductive organs, gizzards, calciferous glands and hearts; and a posterior part rearward of the hindgut comprising a series of rather similar segments.

Earthworms are invertebrates that may be present in very high densities. They are found in litter and soil in all except the coldest regions of the world. They feed on decaying litter and plant residues in soil. Earthworms are extremely important ecosystem engineers, and play a vital role in soil fertility. They are solitary and they are sampled most effectively by hand-sorting soil samples. Earthworms have developed a range of adaptive strategies to live in soil. They can be divided into functional groups according to their feeding habits. They are classed into three main ecological types: epigeic, anecic and endogeic. Those that are pigmented (presence of pigment in skin, generally more concentrated in the dorsal part) are divided into epigeic and anecic earthworms.

Epigeic earthworms live within the litter layers, a changing environment, where they are subject to occasional drought, extreme temperatures and high predator densities. They are generally small (less than 15 cm long on average when adult), homogeneously pigmented (green, blue or reddish) and have rapid movements. Anecic earthworms feed on surface litter that they mix with soil. They spend most of their time in the soil. Anecic earthworms are large (more than 15 cm long when adult). They are pigmented dark green, blue, brown or reddish and the pigmentation is concentrated in the antero-dorsal part of the body. They dig subvertical galleries in the soil. The tail may be flat and enlarged in most common species.

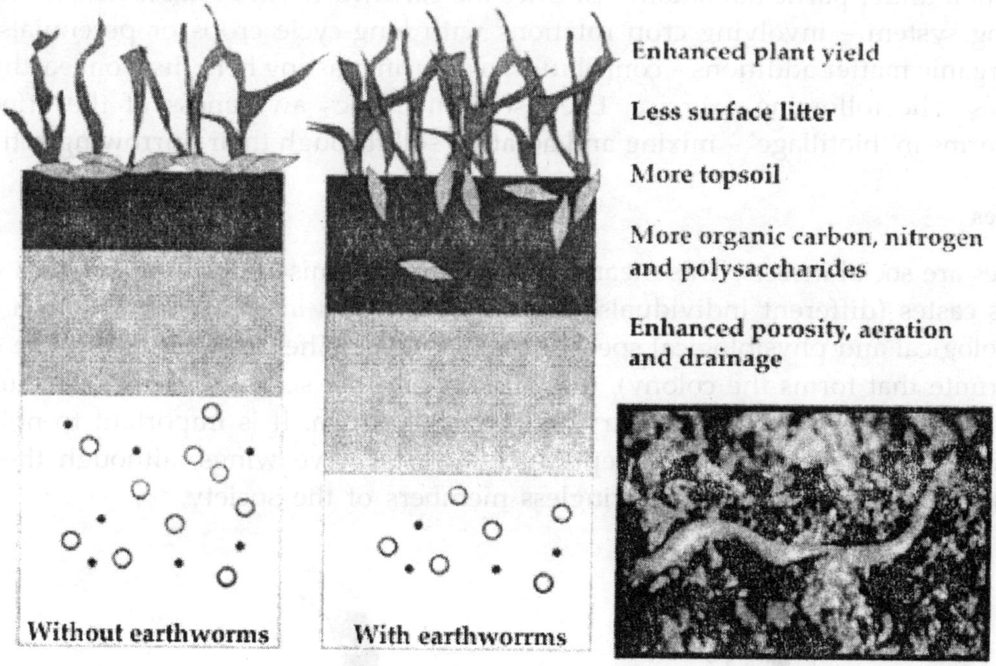

Figure 1. Earthworm—services to soil

The unpigmented earthworms (no pigment in skin, same colour on the dorsal and ventral parts, pink or slightly brown owing to the soil ingested) that live and feed in the soil are called endogeic earthworms. The effects of earthworms in the soil differ according to the ecological category of the species involved. However, in many agricultural systems there may be only one type of earthworm. As a result of their wide range of adaptations, earthworms have diverse functions in the soil. For example, epigeic earthworms are very efficient at making compost but have no impact on soil structure. Anecic earthworms strongly influence soil properties, they build a network of galleries near to the soil surface and deposit casts on the soil surface. Their activity enables the incorporation of large amounts of leaf litter into the soil. Endogeic earthworms are very important in soil structure because of their burrowing activities and their impact on soil aggregation and SOM stabilisation.

Earthworms generally exert beneficial effects on plant growth. However, negative or null effects may be induced in particular situations. The effect on grain yields is also proportional to the earthworm biomass (> 30 g fresh weight), although very high biomasses of a single species of earthworms (e.g. Pontoscolex corethrurus) may inhibit production under particular situations. Once the earthworms are established, a dynamic cropping system – involving crop rotations with long-cycle crops or perennials with good organic matter additions – contributes to securing lasting benefits from earthworm activities. The following exercise, Exercise 5, increases awareness of the effects of earthworms in 'biotillage' – mixing and aerating soil through their burrowing activities.

Termites

Termites are social insects. This means that they are organised, forming colonies where various castes (different individuals with different roles in the colony) with a set of morphological and physiological specialisations coexist. The main castes are: the queen (the termite that forms the colony), the workers and the soldiers . They are abundant and considered a serious pest in dryland areas of Africa. It is important to note that some individuals of the society (reproductive ones) have wings, although they will always be found with the other wingless members of the society.

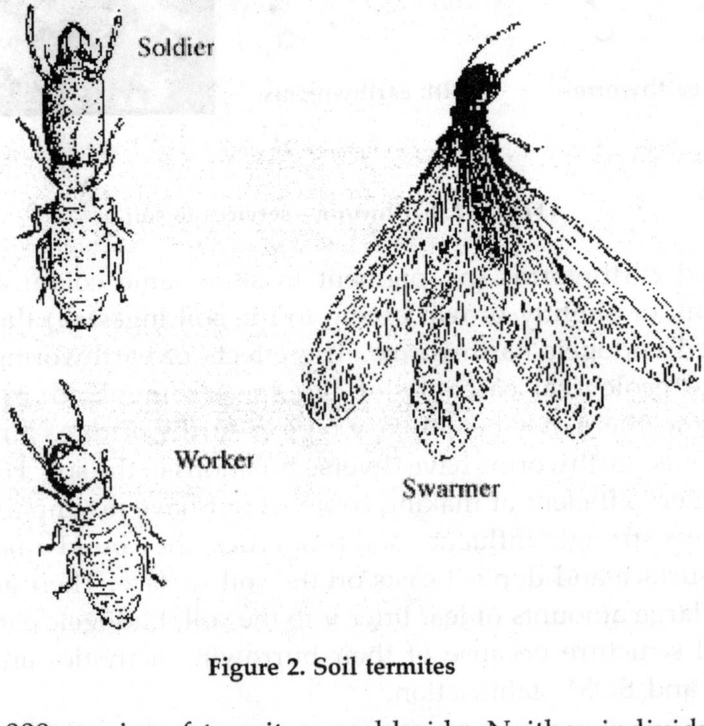

Figure 2. Soil termites

There are about 3 000 species of termites worldwide. Neither individual termites nor colonies normally travel long distances. This is because they are constrained to live within their territorial border or within their food materials. A number of species feed on living plants and some may become serious pests in agricultural systems where dead residues are scarce, e.g. in forest plantations. Most species feed on dead plant materials above or even below the soil surface. These materials include decaying material, e.g. dead foliage of vegetation, woody materials, including roots, seeds and even the faeces of higher animals. There are also soil–wood feeders and soil feeders, this means they ingest a high proportion of mineral material. Their nutrition derives mainly from well-decayed wood and partly-humified SOM. Another group of termites grows fungi in its nests (fungus-growing termites). In general, termites may be separated into five broad groups according to the type of food they ingest:

Grass harvesters: these termites harvest the dead leaves of grasses, which may be stored in their nests; they belong to the families Hodotermitidae and Termitidae. A number of species in this category also collect litter: the tropical Australian species Drepanotermes rubriceps stores a wide range of materials in its low epigeal mounds including leaves of grasses, broad-leaved species and a diversity of seeds and small woody litter. Surface litter feeders: termites that forage for leaf litter, live or dry standing grass stems and small woody items, usually cutting the material before consumption or portage to the nest system. They include some subterranean and mound-building Macrotermitinae as well as certain Nasutitermitinae that forage on the surface of the ground, and at least one lower termite, Hodotermes mossambicus, with a similar habit.

Wood feeders: termites feeding on wood and excavating galleries in larger items of woody litter, which may become colony centres. This group also includes termites having arboreal, subterranean or epigeal nests but feeding elsewhere, and many Macrotermitinae cultivating fungus gardens. Soil-wood feeders: termites feeding in highly decayed wood which has become friable and soil-like, or predominantly within soil under logs or soil plastered on the surface of rotting logs or mixed with rotting leaves.

This group is synonymous with "intermediate feeders", sensu de Souza and Brown, but not the same as the category "rotten-wood feeders" recognised by Collins. Soil feeders (humivores): termites distributed in the soil profile, the organic litter layer and/or epigeal mounds, feeding deliberately on mineral soil, apparently with some degree of selection of silt and clay fractions. Although the ingested material is highly heterogeneous, there are higher proportions of SOM and silica, and lower proportions of recognisable plant tissue than in other groups.

The categories are not mutually exclusive and many species will feed upon at least two sources, especially under unfavourable conditions. Termites are predominantly tropical in distribution. They may reach extremely high densities. Termites are important ecosystem engineers, and may have a similar role to earthworms in promoting soil fertility in tropical systems. They influence: (i) soil porosity and texture through tunnelling, soil ingestion and transport and gallery construction; and (ii) nutrient cycling through the transport, shredding and digestion of organic matter.

Ants

Ants may occur in great numbers in soils and on their surfaces. Ants are effective predators, influencing herbivore populations and plant productivity. As with termites, ants modify soil chemical and physical properties by transporting food and soil materials during feeding and mound and gallery construction. Because of their feeding habits, they are of less importance in regulating processes in the soil than termites or earthworms. In agricultural systems, the leaf-cutting ants of the genera Atta, Acromyrmex and Trachymyrmex are exceptions. They construct very large nests and their harvesting may lead to the incorporation of large amounts of organic matter and nutrients into the soil. In some locations, ants may have an important role in bioturbation and add their effects to those of other groups of soil macrofauna.

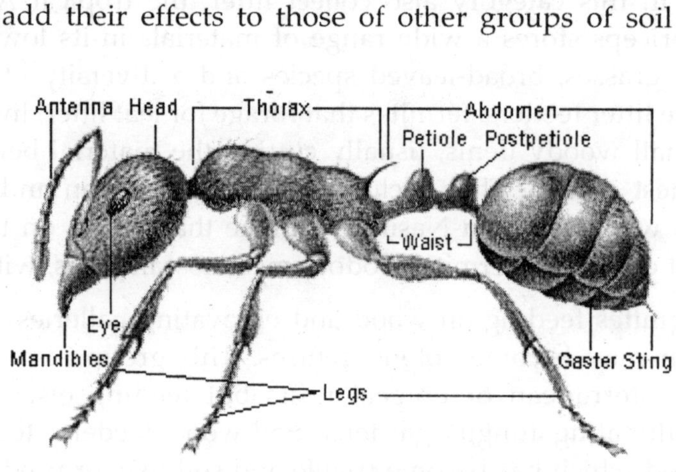

Figure 3. Ant

In agriculture, although generally seen as pests, ants may protect crop herbivores such as honeydew-secreting Homoptera or act directly as herbivores themselves. Their beneficial role as effective predators should also be noted.

Where they are abundant, ants modify the physical structure of the soil through the creation of systems of galleries and chambers. This activity influences soil porosity, aeration, infiltration and drainage and creates habitats for smaller organisms.

The activities of ants can also influence the soil chemistry by increasing the amounts of organic matter, P, K and N in the mounds. The physical changes and the elevated chemical status of many soil materials associated with mounds induce greater mineralisation activity by decomposers and root and mycorrhizal growth.

Other macrofauna such as woodlice, millipedes and some types of litter larvae, which act as litter transformers with an important shredding action on dead plant tissue, and their predators (centipedes, larger arachnids and some other insects) are also key functional groups that have an important influence on soil fertility.

Myriapoda

The group of Myriapoda comprises all soil invertebrates with more than seven pairs of legs. The largest myriapods are divided into two classes: Chilopoda (centipedes) and Diplopoda (millipedes). Millipedes and centipedes are found on the soil surface, in the litter layers, under tree stumps and decaying logs where they can find food and humidity. Most have a limited ability to penetrate the soil. They move through it by displacing it in all directions. Millipedes are generally large- bodied arthropods that feed on plant debris (decomposed wood or leaf materials), leaving behind their numerous droppings that contribute to humus and soil formation.

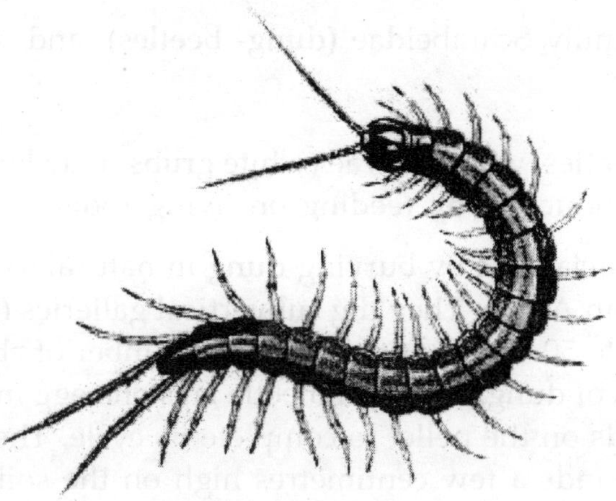

Figure 1. A centipede

Centipedes are generally long, flattened, sometimes large-bodied animals that actively hunt and eat other invertebrates in the litter. However, some very small, unpigmented, blind centipede species (mostly Geophilomorpha) go down into the soil, following galleries and crevices to about 50 cm. They have large jaw-like structures (forcipules) with poison glands at their base (predators).

Fly Larvae

Flies (Diptera) larvae are a very diverse group with an extremely wide range of food sources. They feed on decaying plant material but they tend to be internal feeders on dead plant and animal remains. Sometimes, they can become predominant in soils and may realise an important reduction in litter mass. They live mostly in litter. They are probably most important in cold temperate soils, and can be very abundant in pastures (e.g. Tipulidae). Although often easy to confuse with Coleoptera larvae, they tend to be more slender, less grub-like, have smaller mouthparts, and move more actively in the soil .

Beetles

Beetles (Coleoptera) are the most diverse group of organisms in the world (with perhaps 3–5 million species). They differ widely in size (1–100 mm) and in the ecological roles they have in the soil and the litter. They have a similarly high diversity of feeding habits, and soil beetles can feed on fungi, plant roots, other invertebrates, buried wood, dung, corpses and other rotting organic matter. They have a wide range of feeding habits, being saprophagous, phytophagous or predators. Three groups are of great relevance in agricultural soils:

(i) larvae from the family Scarabeidae (dung- beetles); and

(ii) Curculionidae and

(iii) Melolonthinae beetles, whose larvae (white grubs) may be abundant in grasslands and affect crop production by feeding on living roots.

Dung-beetles play a crucial role by burying dung in natural savannahs and grasslands used for cattle grazing in Africa. They dig subvertical galleries (like anecic earthworms) 10–15 mm wide down to 50–70 cm with a variable number of chambers that are further filled with large pellets of dung. The adult beetle lays one egg in each of these chambers and then the larva feeds on the pellet to complete its cycle. Their presence is generally indicated by small mounds a few centimetres high on the soil surface. The following exercises, Exercises 6 and 7, enables participants to classify soil macrofauna, estimate their abundance and observe their effects on the soil, especially on soil structure, porosity and hence water infiltration.

BIOGENIC STRUCTURES CREATED BY SOIL MACROINVERTEBRATES

Biogenic structures are those structures created biologically by a living organism. They are mainly earthworm casts, termite mounds and ant heaps. The biogenic structures can be deposited on the soil surface and in the soil, and they generally have different physical and chemical properties from the surrounding soil. The colour, size, shape and general aspect of the structures produced by large soil organisms can be described for each species that produces it. The form of the biogenic structure can be likened to simple geometric forms in order to evaluate more easily the volume of soil moved through each type of structure on the soil surface.

Through these structures, the organisms that produce them can modulate the availability or accessibility of one or more resources used by other organisms. Therefore, their activities, including the building of biogenic structures, are capable of modifying the abundance and/or community structure of populations of other organisms without being involved directly in any trophic relationship (e.g. predation, parasitism, mutualism and competition). There are several main groups of biogenic structures that are commonly found in agricultural systems, with different importance and consequences in agro-ecosystems.

Earthworm Casts

Depending on the size of the earthworms that produce them, casts may range from some millimetres to several centimetres in diameter, weighing only a few grams or more than 400 g:

Granular: These casts are very small and are formed by isolated faecal pellets. These casts can be found on the soil surface or in the soil, and are generally produced by epigeic earthworms. Globular: These casts are larger and formed of large aggregates. These are normally produced by endogeic and anecic earthworms. The casts produced by anecic earthworms comprise an accumulation of somewhat isolated, round or oval-shaped pellets (of one to several millimetres in diameter) that may coalesce into "paste-like slurries" that form large structures. Hence, casts are large in size and tower-like, and made of superposed layers of different ages, the older (i.e. dry and hard) located at the base and the more recent (i.e. fresh and soft) on the top. Casts produced by anecic earthworms have a higher proportion of organic matter, especially large particles of plant material and a larger proportion of small mineral components than in the surrounding soil. Both granular and globular casts are normally found in agro-ecosystems in sub-Saharan Africa.

Earthworm Burrows

Earthworms construct burrows or galleries through their movement in the soil matrix. The type and size of the galleries depends on the ecological category of earthworm that is producing it. Anecic earthworms create semi-permanent subvertical galleries, whereas endogeic worms dig rather horizontal burrows. These galleries may be filled with casts that can be split into smaller aggregates by other smaller earthworms or soil organisms. The galleries are cylindrical and their walls coated with cutaneous mucus each time the worm passes through. Soil microorganisms (bacteria) concentrate on the surface of the gallery walls and within the adjacent 2 mm of surrounding soil. This micro-environment comprises less than 3 percent of the total soil volume but contains 5–25 percent of all the soil microflora; it is where some functional groups of bacteria predominate.

Termite Mounds

Termite mounds (termitaria) are among the most conspicuous features in sub- Saharan Africa, especially in savannah landscapes. Termite mounds are of diverse types and they are the epigeal part of a termite nest that originates belowground. Therefore, the termite mounds have at least some part of their structure below the ground surface. In Africa, termites build up half of the biomass of the plains. Their nests may occur in different locations, e.g. within the wood of living or dead trees, in subterranean locations, in other nests formed by other termite species, and by forming epigeal and arboreal nests.

Termites process large quantities of material in their building activities, thereby affecting the soil properties as compared with surrounding soils. Soil texture and structure are strongly modified in the termite mounds. In general, the soil of termite mounds exhibits a higher proportion of fine particles (clay), which termites transport from the deeper to upper soil horizons. Termites that build epigeic domes normally cement soil particles with variable quantities of salivary secretions and excrements rich in organic matter. The enrichment of faecal organic matter explains the differences in concentrations of both C and mineral elements observed between termite mounds and the soil .

Ant Heaps

Because of their feeding habits, ants may be of less importance than termites and earthworms in regulating soil function. The exception to this is the tropical American genus Atta, the leaf- cutting ant. These ants make subterranean nests and their leaf harvesting may lead to enormous incorporations of organic matter and hence nutrients into the soil. Many other ants nest in the soil. In some locations, ants may be important agents of bioturbation. A number of species also concentrate plant nutrients in their nests and the surrounding soils. Ground-dwelling ants, particularly the mound- building ants, can be considered ecosystem engineers in that they modulate the availability of resources and alter the soil and surface environments in ways that affect other organisms.

As with termites, ants also modify soil chemical and physical properties by transporting food and soil materials during feeding and mound and gallery construction. These activities affect soil developmental processes and fertility and may modify the nature and distribution of the vegetation. Where abundant, ants modify the physical structure of the soil through the creation of systems of galleries and chambers. This activity influences soil porosity, aeration, infiltration and drainage, and it also creates habitats for smaller soil organisms. Ant activities can also influence the chemistry of the soil, notably by increasing the amounts of organic matter, P, K and N in the mounds. Many soil materials associated with ant mounds induce greater mineralisation activities by decomposers and heightened root and mycorrhizal growth.

Roots

Although not generally considered soil organisms, they grow mostly within the soil and have wide-ranging, lasting effects on both plant and animal populations aboveground and belowground. Therefore, they are included in soil biota. The rhizosphere is the region of soil that is immediately adjacent to and affected by plant roots. It is a very dynamic environment where plants, soil, microorganisms, nutrients and water meet and interact. The rhizosphere differs from the bulk soil because of the activities of plant roots and their effect on soil organisms.

Roots produce exudates that can help to increase the availability of nutrients in the rhizosphere and they also provide a food source for microorganisms (bacteria). This results in a larger number of microorganisms in the rhizosphere than in the bulk soil. Their presence attracts larger soil organisms that feed upon these microorganisms. The concentration of organisms in the rhizosphere can be up to 500 times higher than in the bulk soil.

An important feature of the rhizosphere concerns the uptake of water and nutrients by plants. Plants take up water and nutrients into their roots. The soil organisms near the rhizosphere influence plant roots because: they alter the movement of C compounds from roots to shoots (translocation); earthworm galleries (burrows) provide an easy pathway for roots to take as they grow through the soil; mycorrhizal associations can increase nutrient uptake by plants; some of them are pathogenic and can attack plant roots, e.g. nematodes. Growing roots also produce important soil aggregation through the production of exudates mixed with clay and other mineral particles. Aggregation in the rhizosphere may alsc result indirectly through the accumulation of faecal pellets of earthworms and other invertebrates that feed in the rhizosphere. This function is very important for preventing soil erosion.

Numerous soil macrofauna groups create biogenic structures (endogeic and epigeic structures) that influence soil processes and structure. Earthworms, termites and ants form the main groups of soil macrofauna recognised as "soil engineers" and the structures that they produce may serve to evaluate their impact on both the soil and other organisms living in it.

Depending on the type of structure considered the impact on soil will be different. The physical structures produced on the soil surface by ecological engineers can be classified into three main categories:

Earthworm casts: very compacted structures, large aggregates, high organic aggregates, low organic C content and assimilable nutrients.

Species producing structures typical of groups 1 and 2 (termites and earthworms) accumulate organic C on the soil surface and probably influence organic matter dynamics and the rate of release of mineral elements assimilable by plants. These structures are characterised by their large size. In contrast, structures in group 3 (termite channels and ant nests) are much smaller.

The production of aggregates with diverse physical-chemical characteristics may result in the efficient regulation of soil structure. For example, in Côte d'Ivoire, the smaller earthworm species break up the casts produced by larger species, thus preventing excessive accumulation on the soil surface. In Carimagua (Colombia), a similar regulation is carried out by termites, it visibly accelerates the degradation of the large casts produces by earthworms. In Amazonian pastures (Manaus, Brazil) the presence of abundant populations of the earthworm species Pontoscolex corethrurus leads to considerable soil compaction because these populations are not associated with species able to break their casts down into much smaller aggregates. The following exercises, Exercises 8 and 9, facilitate the observation of biogenic structures created by soil macrofauna and the estimation of quantities of soil moved and relate soil macrofauna to their ecological roles in the soil.

KEY INDICATOR GROUPS AND OTHER SOIL ORGANISMS

Most soil organisms live in a variety of symbiotic relationships. Symbiotic relationships include: mutualism (both organisms benefit); commensalism (one organism benefits, the other does not but is not harmed); competition; parasitism. These relationships allow many diverse organisms to live in conditions that they could not live in on their own. Together they create substances and recycle materials that create the conditions necessary for life in the soil. Thus, some key indicator groups present these kinds of relationships with other soil organisms. For example, termites are associated closely with specific microbial communities related with termite digestion. Termites also interact with other soil organisms by mutualism, symbiosis, commensalism and predation or parasitism. There is a symbiotic relationship between termites and ants, some termite species profit from the presence in their nests of ants to feed on the residues of dead individual ants.

In some cases, red wood ants have been found in association with earthworms. It is a mutualistic relationship; the surface of the ant nest mound provides a better environment than that of the surrounding soil for earthworms (favourable temperature, moisture and pH, and an abundant food supply) and earthworms prevent the nest mounds from becoming overgrown by moulds and fungi.

Another example of the association of a key indicator group with microorganisms are the fungus-growing ants (which include the leaf-cutting ants). These ants collect various materials and feed them to a symbiotic fungus that lives in their nests. The ants then feed on special nutritional bodies produced by the fungus. This is an example of mutualism. The ants obtain food from the fungus, and the fungus has a place to live protected by the ants from predators and parasites. Large-blue butterfly larvae spend most of their larval stage inside ant nests, either eating ant larvae or being fed by the ants as if they were the ants' own brood (like cuckoos). This is an example of parasitism. The butterflies are dependent on the ants for survival, and have evolved special mechanisms to trick the ants into looking after them.

However, there is another type of relationship between soil organisms that does not rely on trophic interactions but on the biogenic structures produced by the ecosystem engineers. Through their activities, earthworms, termites and ants produce a large variety of macropores (e.g. galleries and chambers) and organo-mineral structures (e.g. earthworm casts, termite mounds, and ant nests) that influence hydraulic properties, macroaggregation and organic matter dynamics in soil.

Through their mechanical and feeding activities, ecosystem engineers modify living conditions (the physical environment and the availability of food resources) for other smaller and less mobile soil organisms, and hence influence their abundance and diversity.

One of the main constraints on the activity of soil organisms is the difficulty of moving in the soil. The mineral soil environment is compact and movement for most species is only possible through the network of pores, galleries and fissures created by the activity of ecosystem engineers.

Biogenic structures, particularly earthworm casts, increase macrofauna density. Some organisms prefer to live inside casts or in the underlying soil because casts have high organic matter content and may represent a valuable food for smaller earthworms and humivorous termites. The biomass of roots is increased locally below casts and this may be beneficial to larvae or rhizophagous Coleoptera. When litter is consumed by earthworms, some changes occur in its composition. These changes attract some litter-dwelling species, such as Isopoda and Diplopoda, that prefer to consume this litter. Small predatory species can find high prey densities of microfauna and mesofauna species taking advantage of earthworm-enhanced living conditions.

Other species may respond to changes in soil structure and to the creation of new specific microhabitats. Macropores that result from earthworm activity can be considered as habitats for some microfauna and mesofauna species. Large numbers of corn-root worm eggs have been found in earthworm burrows. Ants and termites have been observed using galleries as communication ways. Where soil is totally lacking in protection for surface living organisms (i.e. without litter and herbaceous layer), earthworm structures may be used as specific refuges by litter-dwelling arthropods and could help their maintenance and/or rapid recolonisation of the soil surface after perturbations.

Earthworm activities tend to depress nematode populations, especially phytoparasitic species. This effect is the result of changes in the soil environment and the activation of nematophagous fungi.

Relationships between biogenic structures produced by ecosystem engineers and other organisms may be critical for the conservation and dynamics of SOM and the regulation of soil physical properties. For example, when small organisms feed on large and compact earthworm casts, they prevent their excessive accumulation on the soil surface, which otherwise may lead to a superficial soil compaction and affect plant growth negatively. Moreover, they may re-activate organic matter decomposition by making organic resources available to microorganisms that were sequestered in dry casts.

REFERENCES

Bollen, Walter B. 1959. *Microorganisms and Soil Fertility.* Oregon State College. Oregon State Monographs, Studies in Bacteriology, No. 1. 22 p.

Filinger, G. A. 1928. Observations on the habits and control of the garden centipede, *Scutigerella immaculata,* Newport, a pest in greenhouses. J. Econ. Entomol. 2: 357-360.

Goldstein, Jerome. 1998. "Compost suppresses disease in the lab and on the fields". *BioCycle.* November. p. 62–64.

Jahn, GC, JA Litsinger, Y Chen and A Barrion. 2007. *Integrated Pest Management of Rice: Ecological Concepts.* In Ecologically Based Integrated Pest Management (eds. O. Koul and G.W. Cuperus). CAB International Pp. 315-366.

Schneider, R.W. (ed.). 1982. *Suppressive Soils and Plant Disease.* The American Phytopathological Society. St. Paul, MN. 88 p.

4

Identification of Plant Diseases

COMMON PLANT PATHOGENS

Organisms that cause infectious disease include fungi, bacteria, viruses, viroids, virus-like organisms, phytoplasmas, protozoa, nematodes and parasitic plants. Not included are insects, mites, vertebrate or other pests that affect plant health by consumption of plant tissues. Plant pathology also involves the study of the identification, etiology, disease cycle, economic impact, epidemiology, how plant diseases affect humans and animals, pathosystem genetics and management of plant diseases. The "Disease triangle" is a central concept of plant pathology for infectious diseases. It is based on the principle that disease is the result of an interaction between a host, a pathogen, and environment condition.

Fungi

The majority of phytopathogenic fungi belong to the Ascomycetes and the Basidiomycetes. The fungi reproduce both sexually and asexually via the production of spores. These spores may be spread long distances by air or water, or they may be soil borne. Many soil borne spores, normally zoospores and capable of living saprotrophically, carrying out the first part of their lifecycle in the soil. Fungal diseases can be controlled through the use of fungicides in agriculture, however new races of fungi often evolve that are resistant to various fungicides.

Significant fungal plant pathogens:

- Ascomycetes
  - Fusarium spp.
  - Thielaviopsis spp. (Causal agents of: canker rot, black root rot, Thielaviopsis root rot)
  - Verticillium spp.

- Magnaporthe grisea (T.T. Hebert) M.E. Barr; causes blast of rice and gray leaf spot in turfgrasses

- Basidiomycetes
  - Rhizoctonia spp.
  - Phakospora pachyrhizi Sydow; causes Soybean rust
  - Puccinia spp.; causal agents of severe rusts of virtually all cereal grains and cultivated grasses

Oomycetes

The oomycetes are not true fungi but are fungal-like organisms . They include some of the most destructive plant pathogens including the genus Phytophthora which includes the causal agents of potato late blight and sudden oak death. Despite not being closely related to the fungi, the oomycetes have developed very similar infection strategies and so many plant pathologists group them with fungal pathogens.

Significant oomycete plant pathogens:

- Pythium spp.
- Phytophthora spp.; including the causal agent of the Great Irish Famine.

Bacteria

Most bacteria that are associated with plants are actually saprotrophic, and do no harm to the plant itself. However, a small number, around 500 species, are able to cause disease. Bacterial diseases are much more prevalent in sub-tropical and tropical regions of the world.

Most plant pathogenic bacteria are rod shaped (bacilli). In order to be able to colonise the plant they have specific pathogenicity factors. There are 4 main bacterial pathogenicity factors:

1. Cell wall degrading enzymes—used to break down the plant cell wall in order to release the nutrients inside. Used by pathogens such as Erwinia to cause soft rot.
2. Toxins These can be non-host specific, and damage all plants, or host specific and only cause damage on a host plant.
3. Phytohormones—for example Agrobacterium changes the level of Auxin to cause tumours.

4. Exopolysaccharides—these are produced by bacteria and block xylem vessels, often leading to the death of the plant.

Bacteria control the production of pathogenicity factors via quorum sensing.

Significant bacterial plant pathogens:

- Burkholderia
- Proteobacteria
- Xanthomonas spp.
- Pseudomonas spp.
- *Phytoplasmas* ('Mycoplasma-like organisms') and spiroplasmas: Phytoplasma and Spiroplasma are a genre of bacteria that lack cell walls, and are related to the mycoplasmas which are human pathogens. Together they are referred to as the mollicutes. They also tend to have smaller genomes than true bacteria. They are normally transmitted by sap-sucking insects, being transferred into the plants phloem where it reproduces.

Viruses, viroids and virus-like organisms

There are many types of plant virus, and some are even asymptomatic. Normally plant viruses only cause a loss of yield. Therefore it is not economically viable to try to control them, the exception being when they infect perennial species, such as fruit trees.

Most plant viruses have small, single stranded RNA genomes. These genomes may only encode 3 or 4 proteins: a replicase, a coat protein, a movement protein to allow cell to cell movement and sometimes a protein that allows transmission by a vector.

Plant viruses must be transmitted from plant to plant by a vector. This is normally an insect, but some fungi, nematodes and protozoa have been shown to be viral vectors.

Nematodes

Nematodes are small, multicellular wormlike creatures. Many live freely in the soil, but there are some species which parasitize plant roots. They are mostly a problem in tropical and subtropical regions of the world, where they may infect crops. Root knot nematodes have quite a large host range, whereas cyst nematodes tend to only be able to infect a few species. Nematodes are able to cause radical changes in root cells in order to facilitate their lifestyle.

Protozoa

There are a few examples of plant diseases caused by protozoa. They are transmitted as zoospores which are very durable, and may be able to survive in a resting state in the soil for many years. They have also been shown to transmit plant viruses.

When the motile zoospores come into contact with a root hair they produce a plasmodium and invade the roots.

Parasitic plants

Parasitic plants such as mistletoe and dodder are included in the study of phytopathology. Dodder, for example, is used as a conduit for the transmission of viruses or virus-like agents from a host plant to either a plant that is not typically a host or for an agent that is not graft-transmissible.

STEPS IN IDENTIFYING PLANT DISEASES

Step1

Decide if the cause could be a cultural disorder, which occurs when the soil is deficient. A cultural disorder can also be caused by the weather (wind; too much or lack of rain; too much light or not enough.) If the leaves on your plants are turning yellow (chlorosis), this may be due to lack of, or excessive, nutrients in the soil; soil temperature and moisture content; extreme weather conditions; lack of iron or manganese or insufficient light. To give your plants a good beginning, start with healthy soil that has the proper pH level, nutrients and drainage. Make sure that your plant is located in an area where it gets the proper amount of sunlight. Gardening involves a lot of trial and error. If your plants aren't thriving, boost your soil and, if necessary, transplant your flowers to a more suitable location. Compost helps in correcting soil deficiencies and slow-acting fertilizers also help remedy the problem, but don't over-do it when it comes to fertilizing. Read and follow the directions on the container (as hard as that is for some of us.) Next year, you'll know better and won't repeat the same mistakes.

Step 2

Determine if your plants have been infected with a fungal disease. Are there pale patches on the leaves, dead spots or sections? Disease fungi are parasites and basically suck the plants dry. Sometimes all you need to do is remove the affected leaves and that will take care of the problem. If that doesn't do the trick, apply an organic fungicide. Some gardeners recommend using a mixture of 1 level teaspoon of baking soda per 2 quarts of water and spraying this concoction on the plants to get rid of powdery mildew. Others recommend garlic mixed with water to create an effective fungus spray. Plain water, sprayed on the plant, is also thought to be an effective method of controlling mildew.

Mulching helps prevent fungal disease and providing adequate space in between your plants is also a preventative measure. Clean up your garden on a regular basis and get rid of infected plants. Do a thorough cleaning in the fall to prevent a repeat of the problem the next year.

Step 3

Consider that your plants might be infected with a bacterial disease, which is evident when the leaves, stems, branches or tubers rot and there is a terrible odor. You can prevent the spread of a bacterial disease by cleaning your garden tools in disinfectant after using them. Some gardeners use Lysol, while others prefer bleach and water. Wash your hands, too, because you can transmit the disease to a healthy plant. Disease bacteria can wreak havoc on flowering plants by releasing an enzyme that destroys the walls in the leaves, stems and tubers. The bacteria are transmitted via water or transported soil.

Step 4

Find out if your plants are suffering from a viral disease, which can be identified by overall poor plant performance; stunted plants and blooms; disfigured blooms; yellow and green mottling on leaves and stems or puckered leaves and chlorosis (yellowing) If it's a virus, it can't be remedied. Get rid of all of your diseased plants. Do not put them in your compost pile. Wash your hands in bleach afterwards.

STEP5

Prevent all of the common diseases by mulching, watering properly, using organic sprays, controlling insects (i.e., aphids), which carry disease. To avoid spreading fungal disease, don't work with the plants when they're wet. Test your soil to determine what nutrients it lacks and needs.

PROTECTION OF GREENHOUSE FROM DISEASES

A greenhouse, or a controlled-environment structure, is necessary for the production of index plants and for indexing. This structure need not be expensive or elaborate. It should provide light, heat and cooling and be sufficiently well constructed to prevent insect intrusion. An entrance with two doors and a darkened vestibule between is desirable as a preventive measure against insect invasion.

The greenhouse covering can be of glass or plastic. Excellent plants can be grown in a simply designed and inexpensive wooden structure covered with heavy fibreglass and containing a good system for heating and cooling. Modern structures are now made with extruded aluminium framing. In areas where hailstorms are frequent, glass coverings should be avoided, but if used they should be protected with wire mesh. Corrugated fibreglass rather than glass is recommended where hail is a problem, and in many respects is preferable since it may be less expensive, does not break and may be easier to construct and maintain.

The size of the greenhouse will depend upon the amount of indexing and research to be carried out. It may be convenient to have two compartments: a cool room for indexing graft transmissible pathogens which are best expressed in plants grown under relatively cool temperatures (22 to 24°C); and a warm room (up to 34 to 35°C) primarily used for preconditioning vines prior to thermotherapy.

Benches

Benches can be made of wood. concrete, wire mesh, plastic or any satisfactory containersupporting system. If wooden benches are used, it is advisable to paint or spray them with 2 percent copper naphthenate solution, which acts as both a wood preservative and a disinfectant. Wood benches can be placed on concrete blocks or on a metal frame or other foundation at a height of not less than 80 cm from the ground.

Flooring

Flooring can be of concrete with provisions for drainage. However, gravel flooring with a concrete walk is recommended. Gravel of 1 to 2 cm diameter should be spread over the ground about 8 to 10 cm thick. This provides good drainage and aids in maintaining sanitation. The greenhouses should be constructed on a welldrained soil base. If this is not possible, supplementary underfloor drainage tiling should be provided prior to construction.

Containers

Grapevines may be grown in rigid plastic or clay containers of any suitable size. Plastic disposable containers are useful for transplanting rooted indicators immediately after chip-bud grafting, prior to transfer to the field. For prolonged maintenance of vines under greenhouse or screenhouse conditions, large cylindrical or tapered plastic or clay pots can be used. Drawbacks of clay pots are that they accumulate salts, are heavier, are subject to breakage and must be soaked in water and washed after each use.

Temperature Control

It is important to have a recording thermograph in each room. These should be periodically calibrated against two thermometers for accuracy. At the end of each week, when charts are changed, the maximum and minimum temperatures should be recorded in a special book. This provides a record for research and is also a means of noting any abnormal changes, which give warning of heating or cooling unit failure or breakdown.

Supplemental Lighting

During the grapevine vegetating season (April to October) no supplemental lighting is required. It is necessary, however, for growing herbaceous indicators in winter months. In the temperate zone (roughly from 30 to 45° N latitude) the addition of 4 to 5 hours of 2 000 to 2 500 lux at the plant level from October to April is useful. Light sources may be fluorescent tubes or incandescent bulbs.

Heating

Heating can be provided by gas heaters with fans, by steam heating using radiators or by steam pipes placed along the sides of the structure. Heat may also be distributed from gas heaters using supplementary fans blowing the heat through perforated plastic tubes. Most gas heaters are placed inside the structure. However, ethylene released by faulty heaters can be very damaging to plants. If feasible, gas heaters should be placed outside rather than inside the structure, and the warm air circulated by a fen inside the greenhouse, preferably by forcing the air through large-diameter perforated plastic tubes. The construction of a greenhouse or screenhouse is best carried out through local builders, with design and facilities suggested by those in charge of indexing.

Cooling

There are three general methods for cooling a greenhouse: introducing air from the outside when the temperature is cooler than that inside the structure; use of evaporative coolers if the relative humidity is low enough to make such cooling effective; and refrigeration. There may be other innovative methods, such as the doublelayered plastic bubble which acts both as an insulator and as a sandwich through which cool air can be forced. Combinations of any of these methods may be used for economy and efficiency depending upon local conditions.

Air Cooling

The simplest and most economical means of cooling a greenhouse is by bringing in outside air to replace the warm air within. This is best accomplished by using fans and thermostatic control. When the temperature rises the thermostat is activated, the fans turn on and the cooler air is drawn through the greenhouse. A greenhouse designed to utilise the cooling ability of the outside air will save much expensive energy and wear on cooling equipment. Air brought in from outside must be screened or filtered to prevent introduction of insects.

As the temperature rises inside the greenhouse the thermostat will activate the fan, thus bringing outside air into and through the house and forcing the warm air outside. When the temperature increases further the thermostat switches on the water pump, which sends water to the cooling cell to begin evaporation cooling.

Another way of using outside air to cool a greenhouse is to have vents at the peak of the roof. Vents may be activated mechanically, by hand or by a thermostatically controlled motor. When the vents are opened, they permit the warmed inside air to rise and bring in the cooler outside air through filtered vents at the lower sides of the structure. Many problems are associated with this method of air cooling, and though it is present in many older installations, it is generally not recommended for a plant laboratory greenhouse.

Evaporator Coolers

Evaporator coolers are recommended for most greenhouses when humidity during summer months is low. An engineering study should be carried out to calculate the cooling ability of evaporator coolers where humidity is moderate or high during the warm months. Evaporator coolers may prove uneconomical and unsound if the relative humidity is too high. However, in some areas evaporator coolers can be combined with refrigeration for efficient cooling.

Refrigeration

Refrigeration can be used to supplement evaporator coolers where the relative humidity is too high during the warmer months, where extra cooling capacity is needed as a supplement for the plants in a cool indexing room or for cooling small individual rooms. Small plastic chambers can be built inside a large greenhouse to give areas of controlled cooling using refrigeration.

SOIL MIX FOR PLANT GROWTH

Plants for indexing, regardless of whether they are herbaceous or woody. should be of the highest quality. Therefore, the soil mixture with its balanced supply of micro- and macronutrients is of prime importance. The University of California (UC) system for producing healthy container-grown plants was developed by Baker and colleagues, based on the John Innes system of soil mixes developed in England. The system was later modified for growing citrus by the addition of micronutrients to the artificial mixture. This mixture is also suitable for growing grapevines.

Ingredients

The basic soil mixture consists of 50 percent Canadian peat moss and 50 percent fine sand, with macro- and micronutrients added to the mix. Although Canadian peat moss is recommended as the prime ingredient because of its superior nutrient retention and chelating ability, other peats from Europe can be used after appropriate testing. Alternative ingredients such as wood shavings complemented with extra nitrogen, sphagnum mosses, perlite or vermiculite can be tried whenever necessary.

It is recommended that a fine sand or silt, with a particle size ranging from 0.05 to 0.5 mm, be used. Beach sand should be avoided. Fine sand can be found in rivers, in wind-blown deposits or as the fine silt separated out as waste material from a sand and gravel processing pit. A quick and simple test for determining the presence of clay in a sand source is to shake a sample of the test soil in a jar with water. If the sand settles fairly rapidly and the water remains relatively clear, it is satisfactory. If clay is present, the water will have a muddy appearance, and that source of sand should preferably not be used. The sand should be inert and preferably siliceous. Calcareous and limestone sands should be avoided since they may affect the pH. If a high-grade silicate sand is not available, consideration should be given to a substitute mix of peat, vermiculite and perlite in a ratio of 2:1:1 or 1:1:1. The objective is to obtain an artificial mixture that is consistently reproducible, will absorb and release macro- and micronutrients and will maintain pH of the drainage water at 5.5 to 6.5.

The ingredients can be mixed together with a shovel on a flat concrete surface.. However, a small or medium-sized electric or gasolinepowered concrete mixer is the preferred mixing device. In this case the procedure is as follows:

- A specific number of uniform. standard shovelscoops of soil, peat and wood shavings, or other substitutes for part of the peat moss, are counted and shovelled into the apron of the concrete mixer.
- A weighed quantity of macronutrients, i.e. phosphate, calcium and magnesium, is sprinkled on top of the unmixed ingredients in the apron.
- The soil ingredients plus macronutrients are then dumped into the concrete mixer and thoroughly tumbled. The micronutrients, pre-weighed and mixed together in a package, are first dissolved in a container of water and then poured into the turning mixer.
- A small quantity of water can be added to the soil while the mixer is turning to bring the soil mixture to a friable, moist level. This may be necessary if the soil or peat is too dry.

- After about 20 minutes of tumbling and mixing, the soil is emptied from the mixer into a trailer fitted on the bottom with 20 mm galvanised steam pipes with 5-mm holes located at the bottom of the pipes, spaced 15 to 20 cm apart. The pipes are spaced 15 cm apart. The trailer top is covered with a cloth tarpaulin. Containers and flats may be placed on top of the soil in the trailer before covering, or steamed separately.
- The soil mixture is then steamed. The steaming time will depend on the quantity of steam produced, which is proportional to the size and capacity of the boiler. A good general criterion is to continue to steam for about 15 minutes after the steam billows the covering tarpaulin. Soil thermometers placed in the corners of the trailer are helpful to judge the correct period for steaming. One minute at 100°C or 10 minutes at 83°C is usually sufficient for controlling soilborne pathogens. Steaming has never been found toxic or harmful to plants grown in UC soil mix with the above composition.

Fertilisation

The initial mix contains both macro- and micronutrients added during mixing. The micronutrients are tied up in the peat moss. The peat, which acts as a chelating agent, releases sufficient small amounts of micronutrients to the plants for up to one to two years.

Liquid fertiliser is applied with each watering using a proportioning device. There are a number of such devices, which inject fertilisers in proportion to the water used. An effective, simple and very inexpensive device is a Venturitype siphon. Before use the siphon should be calibrated, for many such devices vary considerably from the advertised ratio of concentrate to water as printed in the instructions. To calibrate the siphon, put a measured amount of water (500 or 1 000 ml) in a graduated cylinder, then place the suction end of the siphon into the cylinder and measure the final amount of water exiting from the hose. Allow the water to fill a container until the 500 or 1 000 ml of measured liquid is siphoned up. Then measure the water in the container and convert the results to a ratio.

Another device that injects a given quantity of liquid fertiliser into the water system at a uniform rate in direct proportion to the water flow is the Smith Measuremix proportioner. This is a highly reliable, precision instrument. However, it should be calibrated in the same manner as the Venturi siphon.

A liquid fertiliser formula based on that given by Nauer, Roistacher and Labanauskas is: 9 parts ammonium nitrate + 3.75 parts calcium nitrate + 2.75 parts potassium nitrate or potassium chloride. This fertiliser should be we 11 mixed and applied at the rate of 67.5 g of mixture to 100 litres of water.

With the UC system of soil mix, the soil should be fertilised directly after mixing and before and immediately after planting since the basic mix contains no nitrogen or potassium. As a general practice, potted plants in a UC system need to be watered periodically with enough volume to flush out any accumulated salts, thereby preventing salinity buildup. It is important that the soil not be filled to the top of the container; a space of 2 to 3 cm should be left between the top of the container and the soil level. This will allow a sufficient volume of water to flush the soil in the container adequately.

MANAGEMENT OF DISEASES

Quarantine

Wherein a diseased patch of vegetation or individual plants are isolated from other, healthy growth. Specimens may be destroyed or relocated into a greenhouse for treatment/study. Another option is to avoid introduction of harmful non-native organisms by controlling all human traffic and activity although legislation and enforcement are key in order to ensure lasting effectiveness.

Cultural

Farming in some societies is kept on a small scale, tended by peoples whose culture includes farming traditions going back to ancient times. Plants that are intently monitored often benefit not only from active external protection, but a greater overall vigor as well. While primitive in the sense of being the most labor-intensive solution by far, where practical or necessary it is more than adequate.

Plant resistance

Sophisticated agricultural developments now allow growers to choose from among systematically cross-bred species to ensure the greatest hardiness in their crops, as suited for a particular region's pathological profile. Breeding practices have been perfected over centuries, but with the advent of genetic manipulation even finer control of a crop's immunity traits is possible.

Chemical

Many natural and synthetic compounds exist that could be employed to combat the above threats. This method works by directly eliminating disease-causing organisms or curbing their spread; however it has been shown to have too broad an effect, typically, to be good for the local ecosystem.

Biological

Crop rotation may be an effective means to prevent a parasitic population from becoming well established, as an organism affecting leaves would be starved when the leafy crop is replaced by a tuberous type, etc. Other means to undermine parasites without attacking them directly may exist.

Integrated

The use of two or more of these methods in combination offers a higher chance of effectiveness.

REFERENCES

Campbell, Robert N., and Arthur S. Greathead. 1990. "Control of clubroot of crucifers by liming". p. 90-101. In: A.W. Engelhard (ed.). *Management of Diseases with Macro- and Microelements.* APS Press American Phytopathological Society, St. Paul, MN.

Engelhard A.W. (ed.). *Soilborne Plant Pathogens: Management of Diseases with Macro- and Microelements* American Phytopathological Society. 217 p.

Johnston, S.A., and P.J. Nitzche. "Rotation periods suggested to help control vegetable diseases". New Jersey Extension Service. 1 p.

Woltz, S.S., and J.P. Jones. 1973. "Tomato Fusarium wilt control by adjustments in soil fertility" Proceedings of the Florida State Horticulture Society. Volume 86. p. 157–159.

5

Nematode Pests of Plants

Nematodes are microscopic roundworms that live in many habitats. At least 2 500 species of plant-parasitic nematodes have been described, characterised by the presence of a stylet, which is used for penetration of host plant tissue. Most attack roots and underground parts of plants, but some are able to feed on leaves and flowers. Plant-parasitic nematodes are of great economic importance. However, because most of them live in the soil, they represent one of the most difficult pest problems to identify, demonstrate and control. Their effects are commonly underestimated by farmers, agronomists and pest management consultants, but it has been estimated that some 10 percent of world crop production is lost as a result of plant nematode damage.

Nematodes have successfully adapted to nearly every ecological niche from marine to fresh water, from the polar regions to the tropics, as well as the highest to the lowest of elevations. They are ubiquitous in freshwater, marine, and terrestrial environments, where they often outnumber other animals in both individual and species counts, and are found in locations as diverse as Antarctica and oceanic trenches. They represent, for example, 90% of all life on the seafloor of the Earth. The many parasitic forms include pathogens in most plants, animals, and also in humans. Some nematodes can undergo cryptobiosis.

Plant parasitic nematodes include several groups causing severe crop losses. The most common genera are Aphelenchoides (foliar nematodes), Ditylenchus, Globodera (potato cyst nematodes), Heterodera (soybean cyst nematodes), Longidorus, Meloidogyne (root-knot nematodes), Nacobbus, Pratylenchus (lesion nematodes), Trichodorus and Xiphinema (dagger nematodes). Several phytoparasitic nematode species cause histological damages to roots, including the formation of visible galls (e.g. by root-knot nematodes) which are useful characters for their diagnostic in the field. Some nematode species transmit plant viruses through their feeding activity on roots. One of them is Xiphinema index, vector of GFLV (Grapevine Fanleaf Virus), an important disease of grapes.

Other nematodes attack bark and forest trees. The most important representative of this group is Bursaphelenchus xylophilus, the pine wood nematode, present in Asia and America and recently discovered in Europe.

Although many nematodes have been found associated with small-grained cereals, only a few of them are considered economically important. Those of importance include:

(i) cereal cyst nematodes, *Heterodera* spp.;

(ii) root lesion nematodes, *Pratylenchus* spp.;

(iii) root knot nematodes, *Meloidogyne* spp.;

(iv) seed gall nematode, *Anguina tritici*; and

(v) stem nematode, *Ditylenchus dipsaci*.

Management of nematodes may be approached by using a combination of methods in an integrated pest management system or may involve only one of these methods. Some of the most commonly practised methods are including crop rotation, the use of resistant and tolerant cultivars, cultural practices and chemicals.

It is important to stress that the most appropriate control method will be determined by the nematode involved and the economic feasibility of implementing the possible control(s). The purpose of this subject is to provide an insight into the economically important nematodes on small grains, their currently known distribution and damage potential, and the management options that exist for their control.

CEREAL CYST NEMATODES

Distribution

The cereal cyst nematodes, *Heterodera* spp., are a group of several closely related species and are considered to be one of the most important groups of plant-parasitic nematodes on a worldwide basis. The most commonly recorded species of economic importance on cereals is *H. avenae*, which has been detected in many countries, including Australia, Canada, Israel, South Africa, Japan and most European countries, as well as India and countries within North Africa and West Asia, including Morocco, Tunisia, Pakistan and Libya, and recently Algeria and Saudi Arabia. Although its distribution is global, much of the research has been confined to Europe, Canada, Australia and India.

Heterodera avenae is the principal species on temperate cereals, while another important cereal species, *H. latipons*, is essentially only Mediterranean in distribution, being found in Syria, Israel, Cyprus, Italy and Libya. However, it is also known to occur in northern Europe. Other *Heterodera* species known to be of importance to cereals include: *H. hordecalis* in Sweden, Germany and the United Kingdom; *H. zeae*, which is found in India, Pakistan and Iraq; *H. filipjevi* in Russia and Turkey; and various others, including *H. mani*, *H. bifenestra* and *H. pakistanensis*, and an unrelated species of cyst nematode, *Punctodera punctata*.

Other cyst nematode species have been found on cereals, but they have not been shown to be economically important. Most of these species are difficult to differentiate easily and require a strong taxonomic under-standing of morphological traits of cysts or juveniles. Recent molecular techniques, such as random fragment length polymorphism (RFLP) of the ribosomal DNA, have enabled solid taxonomic differentiation among several entities of the cereal cyst nematode complex.

Biology

The host range of *H. avenae* is restricted to graminaceous plants. There is sexual dimorphism with the male remaining worm-like, whereas the female becomes lemon-shaped and spends its life inside or attached to the root. The adult white female is clearly visible on roots with the swollen body, about 1 mm across, protruding from the root surface. Eggs are retained within the female's body, and after the female has died, the body wall hardens to a resistant brown cyst, which protects the eggs and juveniles. The eggs within the cyst remain viable for several years. *Heterodera avenae* has only one generation per year, with the hatch of eggs determined largely by temperature.

The symptoms produced on the roots are different dependent on the host. Wheat attacked by *H. avenae* shows increased root production such that the roots have a 'bushy knotted' appearance usually with several females visible at each knot as illustrated in Plate 55. Oat roots are shortened and thickened, while barley roots appear less affected. Other species of *Heterodera* also appear to produce host-specific symptoms on the roots of cereals. For example, in Israel *H. latipons* did not produce knotted roots as *H. avenae*. Above-ground symptoms of *H. avenae* appear early in the season as pale green patches of plants with fewer tillers. Patches may vary in size from 1 m^2 to 100 m^2 or more.

In France, successful detection of *H. avenae* in wheat fields was achieved with the use of radiothermometry. It is possible that this technique could be extended to thermography, which could improve the detection of cereal cyst nematode attacks in large areas.

Heterodera avenae is the best known species, but is polymorphous with many pathotypes. The induction or suppression of dormancy by different temperatures regulates the hatching of *H. avenae* juveniles. In Mediterranean climates, the diapause is obligate and durable, acting when the climate is hot and dry and being suppressed when the soil temperature falls and moisture rises.

The diapause requirements in other climates with *Heterodera* species are less well understood but they are essential to understanding the biology and control of those species. To date, the pathotypes of *H. avenae* have been recognised with the test developed by Andersen and Andersen designated

Economic importance

Heterodera avenae has been associated with economic levels of damage exclusively in light soils. However, it can cause economic damage irrespective of soil type when the intensity of cereal cropping exceeds a certain limit. Yield losses due to this nematode are: 15 to 20 percent on wheat in Pakistan; 40 to 92 percent on wheat and 17 to 77 percent on barley in Saudi Arabia; and 20 percent on barley and 23 to 50 percent on wheat in Australia.

Recent studies by Scholz implicate yield loss with both barley and durum wheat with *H. latipons*. Also *H. avenae* and *H. zeae* are major pests of wheat and barley in Pakistan.

In India, *H. zeae* is considered to be one of the most economically important nematodes attacking cereals. *Heterodera avenae* has been associated with severe diseases present in India known as *molya*, but it only occurs on temperate cereals, such as barley and wheat, while tropical cereals, such as sorghum and maize, are non-hosts. In the northwestern part of India, between four- and sixteen-fold increases in yield of wheat and barley have been obtained after nematicide treatments.

Staggering annual yield losses of 3 million pounds sterling in Europe, 72 million Australian dollars in Australia and 9 million US dollars in India have been calculated as being caused by *H. avenae*. The losses in Australia are now greatly reduced due to control of the disease with resistant and tolerant cultivars.

Little is known about the economic importance of the species *H. latipons*, even though it was first described in 1969. Field studies in Cyprus indicated a 50 percent yield loss on barley. Because the cysts are similar in size and shape, it is possible that previous findings of this recently described nematode species have erroneously been attributed to the economically important *H. avenae*.

In West Asia and North Africa, *H. latipons* has been found on wheat and barley in four countries. It has also recently been confirmed in Turkey. It has also been reported from several Mediterranean countries associated with the poor growth of wheat.

Unfortunately, this nematode has not been studied in detail, and information on its host range, biology and pathogenicity is scarce; none-theless, it is suspected to be an important constraint on barley and durum wheat production in temperate semi-arid regions. Other cyst nematodes, such as *P. punctata* and *H. hordecalis*, have been described from roots of cereals in several countries, but their distribution and economic importance is unknown.

Control

One of the most efficient methods of controlling *H. avenae* is with grass-free rotations using non-host crops. In long-term experiments, non-host or resistant cereal frequencies of 50 percent (80 percent in lighter soils) keep populations below damaging thresholds. Clean fallow and/or deep summer ploughing reduce the population density of the nematode but are not always environmentally sound.

Cultivar resistance is considered one of the best methods for nematode control and has been found to be successful in several countries such as Australia, Sweden and France on a farm scale. However, it has also been observed that the use of resistance, especially derived from single dominant genes, may cause a disequilibrium in the biological communities and possibly ecological replacement with other nematodes, such as *Pratylenchus*.

Another potential concern is the breakdown of resistance sources with repeated use. This has occurred in France with the resistant oat cultivar Panema and the appearance of a new *H. avenae* pathotype. In order for cultivar resistance to be effective and durable, a sufficient understanding of the number of species and pathotypes within species is essential.

The International Cereal Test Assortment for Defining Cereal Cyst Nematode Pathotypes offers classification of pathotype variation; pathotypes from Australia and India are often distinct from those in Europe.

Although useful, a pathotype scheme for a species complex based on interaction with three cereal genera will not easily describe extensive variation in virulence. Furthermore, to date there are few molecular or other diagnostic methods that can provide consistent and reliable pathotype and pathogenicity differentiation.

Some resistant cultivars simultaneously reduce populations of several European pathotypes. Since this review, developments have found additional *Triticum* accessions that appear to possess high degrees of resistance to a broad array of *Heterodera* species and pathotypes. Molecular technology has also been applied to identify markers for various cereal cyst nematode resistance genes using techniques such as RFLP and polymerase chain reaction (PCR) in both barley and wheat.

Furthermore, many of the wild grass relatives have been introgressed into a hexaploid wheat background for breeding purposes. Many of these have had molecular work applied to identify the location and the possibility to produce markers to the known gene(s).

Some of these markers are actively being implemented in marker-assisted selection and pyramiding of gene resistance in Australian cereal breeding programmes against *H. avenae,* pathotype Ha13. This is an example where there is sufficient under-standing of the biology of the pathogen and genetic control of the resistance so that both conventional breeding and the modern tools of molecular biology can be combined to the advancement of controlling this disease. Such potential exists for other nematodes, but will require a similar understanding and combining of related skill base.

The utilisation of these identified sources, and possibly of other as yet unidentified sources of resistance, is country-specific and dependent on the number and types of *Heterodera* species and pathotypes that need to be controlled. Many developing countries unfortunately have limited resources and/or expertise to establish this information, and current control methods are based on understanding the response of local cultivars to the pathogen(s). For example, in Israel all locally grown wheat and barley cultivars tested against *H. avenae* and *H. latipons* are excellent hosts. However, the oat cultivars tested were extremely poor hosts to *H. avenae* but good hosts to *H. latipons*.

In Mediterranean countries, such as Algeria, Spain, Israel and southern France, oats appeared generally to be a poor host for *H. avenae,* in comparison to northern Europe where they are considered to be a good host, suggesting the possibility that the nematode has developed host race types. In order to make the best use of existing research findings, greater collaboration between research institutions and countries where the nematode is considered important is essential.

An excellent example of this is the most recent report by Rivoal *et al.*, which offers a great start to unravelling the complexity of *Heterodera* populations and the existing knowledge of resistant sources and their possible uses in controlling the cereal cyst nematode in different regions of the world.

The use of chemical fumigants and nematicides, although proven effective in experimental fields in many countries, is not an economically feasible option for most farmers. Application of nematicides for the control of *H. avenae* on wheat has resulted in 50 to 75 percent yield increases in Pakistan, but their use is not feasible on a commercial scale.

The deployment of biological control agents is not yet an option, but natural biological control has been found to operate in some circumstances. The fungi *Nematophora gynophila* and *Verticillium chlamydosporium* have been associated with reduction and suppression of *H. avenae* populations under intensive cereals in the

United Kingdom, and similar suppression may occur in other regions with similar climates.

ROOT LESION NEMATODES

Distribution

The genus *Pratylenchus* is a large group with many species affecting both monocots and dicots. Many of the species are morphologically similar, which makes them difficult to identify. At least eight species of lesion nematodes have been recorded for small grains. Of these, four species (*P. thornei, P. crenatus, P. neglectus* and *P. penetrans*) have a worldwide distribution, especially in the temperate zones. *Pratylenchus crenatus* is more common in light soils, *P. neglectus* in loamy soils and *P. thornei* in heavier soil types. However, the work of Nicol suggests that both *P. thornei* and *P. neglectus* can occur in a range of soil types, and mixtures of the two species are not uncommon in southern Australia.

Pratylenchus thornei is the most studied species on wheat and is a known parasite of cereals worldwide, being found in Syria, former Yugoslavia, Mexico and Australia, Canada, Israel, Morocco, Pakistan and India, Algeria and Italy. Unfortunately, very little is known about the economic importance and distribution of the other species on cereals.

Biology

Pratylenchus species are polycyclic, polyphagous migratory root endoparasites that are not confined to fixed places for their development and reproduction. Eggs are laid in the soil or inside plant roots. The nematode invades the tissues of the plant root, migrating and feeding inside the root causing characteristic dark brown or black lesions on the root surface, hence its common name. Extensive lesioning, cortical degradation and reduction in both seminal and lateral root systems is seen with increasing nematode density, as illustrated in Plate 56. Secondary attack by fungi frequently occurs at these lesions. The life cycle is variable between species and environment and ranges from 45 to 65 days. Above-ground symptoms of *Pratylenchus* on cereals, as with other cereal root nematodes, is non-specific where infected plants appear stunted and unthrifty, sometimes with reduced numbers of tillers and yellowed lower leaves.

Economic Importance

The most studied of these species on wheat is *P. thornei* and, some-what less so, *P. neglectus* and *P. penetrans*. *Pratylenchus thornei* is considered the most economically important species in at least three countries; yield losses on wheat have been reported between 38 and 85 percent in Australia, 32 percent in Mexico and 70 percent in Israel. *Pratylenchus thornei* appears to be associated with regions experiencing a Mediterranean climate. It is highly probable, given the distribution of this nematode, that similar losses may also be occurring in many other countries, however this has not been studied.

The other species of lesion nematodes where yield loss studies have been conducted (*P. neglectus* and *P. penetrans*) are not recognised as having a global distribution on cereals, and the current yield loss studies would suggest that the damage potential of these nematodes is not as great as that of *P. thornei*. In Australia, losses on wheat with *P. neglectus* ranged from 16 to 23 percent, while in Canada *P. penetrans* losses were 10 to 19 percent. Yield loss work by Vanstone *et al*. in the field where both *P. thornei* and *P. neglectus* were present indicates losses between 56 and 74 percent on wheat.

Recent studies by Sikora have identified *P. neglectus* and *P. penetrans* in addition to *P. thornei* on wheat and barley in North Africa and all of these plus *P. zeae* in West Asia. Further work is necessary to determine the significance of these species in these regions.

Control

Unlike cereal cyst nematode, no commercially available sources of cereal resistance to *P. thornei* are available, although sources of tolerance have been used by cereal farmers in northern Australia for several years. Work by Thompson and Clewett and Nicol *et al*. identified wheat lines that have proven field resistance, and work is continuing to breed this resistance into suitable backgrounds. Recent work by Thompson and Haak identified 29 accessions from the D-genome donor to wheat, *Aegilops tauschii*, suggesting there is future potential for gene introgression.

Some of this material also contained the *Cre3* and other different unidentified sources of cereal cyst nematode resistance genes conferring resistance to some cereal cyst nematode pathotypes. As with cereal cyst nematode, molecular biology is also being used to investigate genetic control and location, followed by the identification of markers for resistance to both *P. thornei* and *P. neglectus*. Recent work with Australian germplasm referred to by McIntosh *et al*. reports the gene *Rlnn1* on chromosome 7AL effective against *P. neglectus*, and two quantitative trait loci on chromosomes 2BS and 6DS have been found for *P. thornei*. No commercial sources of resistance are currently available for other species of *Pratylenchus* that attack cereals.

The use of crop rotation is a limited option for root lesion nematodes due to the polyphagous nature of the nematode. Little is understood about the potential role of crop rotation in controlling these nematodes, although some field and laboratory work has been undertaken to better understand the ability of both *P. thornei* and *P. neglectus* to utilise cereals and leguminous crops as hosts. It is possible, depending on crop rotation patterns and the population dynamics of the nematodes, that resistant cultivars of cereals alone may not be sufficient to maintain the nematode below economic levels of damage.

As with other nematodes, chemical control, although in most cases effective against root lesion nematodes, is not economically viable with cereal crops. Cultural methods offer some control options, but are of limited effectiveness; in order to be of major significance, they need to be integrated with other control measures. Di Vito *et al.* found that mulching fields with polyethylene film for six to eight weeks suppressed *P. thornei* populations by 50 percent. Van Gundy *et al.* found that delaying the sowing of winter irrigated wheat by one month in Mexico gave maximum yields. In Australia, cultivation reduced populations of *P. thornei*.

ROOT KNOT NEMATODES

Distribution

Several *Meloidogyne* spp. are known to attack cereals and tend to favour light soils and warm temperatures. Several species attack Poaceae in cool climates, including *M. artiellia*, *M. chitwoodi*, *M. naasi*, *M. microtyla* and *M. ottersoni*. In warm climates, *M. graminicola*, *M. graminis*, *M. kikuyensis* and *M. spartinae* are important. In tropical and subtropical areas, *M. incognita*, *M. javanica* and *M. arenaria* are all known to attack cereal crops.

To date, only *M. naasi* and *M. artiellia* have been shown to cause significant damage to wheat and barley in the winter growing seasos. The most important and most studied species of the root knot nematodes on cereals worldwide are described below. There is little information on most other species, many of which are of unknown importance.

Meloidogyne naasi is reported from the United Kingdom, Belgium, the Netherlands, France, Germany, former Yugoslavia, Iran, the United States and the former Soviet Union, occurring mostly in temperate climates. However, it has also been found in Mediterranean areas on barley in the Maltese islands and in New Zealand and Chile on small grains.

It is probably the most important root knot nematode affecting grain in most European countries in contrast to the United States. *Meloidogyne naasi* does not appear to be widespread in temperate semi-arid regions, such as West Asia and North Africa.

Meloidogyne naasi is a polyphagous nematode, reproducing on at least 100 species of plants including barley, wheat, rye, sugar beet, onion and several broadleaf and monocot weeds. However, Poaceae are considered to be better hosts. In Europe, oats are a poor host compared with other cereals, whereas in the United States, oats are an excellent host of *M. naasi.* Host races of *M. naasi* have been identified in the United States by using differential hosts, which makes controlling this nematode more difficult.

Another species of root knot nematode that attacks cereals is *M. artiellia,* which has a wide host range including crucifers, cereals and legumes. It is known to reproduce well on cereals and severely damage legumes. This nematode is chiefly known from Mediterranean Europe in Italy, France, Greece and Spain, but also in West Asia, Syria, Israel and western Siberia.

Meloidogyne chitwoodi is a pest on cereals in the Pacific Northwest of the United States and is also found in Mexico, South Africa and Australia. Many cereals, including wheat, oats, barley and maize, and a number of dicots are known to be hosts. *Meloidogyne graminis* is not known to be widely distributed, being limited to the southern United States where it is associated with cereals and more often turf grasses.

Biology

The young juveniles of *M. naasi* invade the roots of cereals within one to one and one-half months of germination, after which small galls on the root tips can be observed. *Meloidogyne naasi* generally has one generation per season. The juveniles develop, and the females become almost spherical in shape. Females deposit eggs in an egg sac. They usually appear eight to ten weeks after sowing and are found embedded in the gall tissue.

Large galls may contain 100 or more egglaying females. Later in the season, galling of the roots, especially the root tips, is common. Galls are typically curved, horseshoe or spiral-shaped. The egg masses in galls survive in the soil. Eggs have a diapause, broken by increasing temperature after a cool period. In warmer regions on perennial or volunteer grass hosts, more than one generation per season is possible.

Symptoms of *M. naasi* attack closely resemble those caused by *H. avenae,* with patches of poorly growing, yellowing plants that may vary in size from a few square metres to larger areas. Other root knot nematodes attacking cereals are suspected to produce similar symptoms, but most are much less studied than *M. naasi.*

Economic Importance

Information on the economic importance of root knot nematodes on cereals is limited to a few studied species. *Meloidogyne naasi* can seriously affect wheat yield in Chile and Europe. On barley, it has been known to cause up to 75 percent yield loss in California, United States. It is also associated with yield loss on barley in France, Belgium and the United Kingdom. Severe losses can occur with entire crops of spring barley lost in the Netherlands and France. *Meloidogyne naasi* damage is not known to be widespread in temperate, semi-arid regions, but rather has been associated with wet and/or over-compacted soils.

Damage to wheat by *M. artiellia* is known from Greece, southern Israel and Italy. In Italy, 90 percent yield losses on wheat have been recorded. *Meloidogyne chitwoodi*, an important pathogen of potatoes, also damages cereals in Utah, United States, and Mexico. In controlled laboratory studies, *M. incognita* and *M. javanica* have been shown to reduce plant growth of wheat, and *M. incognita* is a known field problem on wheat in northwestern India.

Control

Control methods for root knot nematodes have only been investigated in detail for the known economically important species *M. naasi*. Partial resistance was found in barley and also in *Ae. tauschii* and *T. monococcum*, while full resistance was identified with *H. chilense*, *H. jabatum*, *Ae. umbellulatum* and *Ae. variabile*. Cultural management options for *M. naasi* include rotations using poor or non-host crops and the use of fallow during the hatching period.

SEED GALL NEMATODES

Distribution

Seed gall nematode (*Anguina tritici*), commonly known as ear cockle, is frequently found on small grain cereals where farm-saved seed is sown without the use of modern cleaning systems. Cereals are infected throughout West Asia and North Africa, the Indian subcontinent, China, parts of Eastern Europe, Iraq and Pakistan. It has also been reported from most European countries, the Russian Federation, Australia, New Zealand, Egypt, Brazil and several areas in the United States.

Biology

The nematode is spread in galled or 'cockled' seeds when infected seed is sown. A single gall may contain over 10 000 dormant juveniles. Once sown, the galls take up water, and the juveniles emerge and remain between the leaves of the growing plant. The primary leaves become twisted and distorted, and the plant may die from a heavy attack.

In growing seedlings, the juveniles are carried upward towards the growing point of the plant, and when the ear is formed, the flower head is invaded by the juveniles. As a result, the ovules and other flowering parts of the plant are transmuted into galls or cockles. Inside the galls, the nematodes mature, and the females lay thousands of eggs from which the juveniles hatch and remain dormant in the seed. The nematode is favoured by wet and cool weather. Symptoms of *A. tritici* attack may be indicated by small and dying plants with the leaves generally twisted due to nematode infection.

The attacked ears are easily recognised by their smaller size and darkened colour compared with normal seeds, but the infected seeds may be easily confused with common bunt (*Tilletia tritici*). Under dry conditions, the juveniles may survive for decades. The nematode is also associated with a bacterium, *Corynebacterium michiganense* pv. *tritici*, which causes yellow ear rot. The economic loss associated with this combination is increased because of the lower price for infected grain.

Economic Importance

Worldwide, wheat, barley and rye are commonly attacked, but barley is less attacked in India. In Iraq, seed gall is an important pest on wheat with infection ranging from 0.03 to 22.9 percent and causing yield losses up to 30 percent. Barley is also attacked in Iraq but with an isolate that does not affect wheat.

In Pakistan, seed gall is a known pest on wheat and barley and is found in nearly all parts of the country, causing yield losses of 2 to 3 percent; association with the bacterium produces serious yield losses on wheat. In China, Chu found yield losses between 10 and 30 percent on wheat.

Control

Seed gall can easily be controlled through seed hygiene: sowing clean, non-infected seed obtained by using certified seed or by cleaning infected seed either with modern seed cleaning techniques or by sieving and freshwater flotation. Although seed gall has been eradicated from the Western Hemisphere through the adoption of this approach, it remains a problem on the Indian subcontinent, in West Asia and to some extent in China.

For countries where hygiene practices are difficult to implement, host resistance and crop rotation offer some control of seed gall. Resistance to *A. tritici* has been identified in Iraq in both wheat and barley and in Pakistan and is currently being sought in India. In Iraq, laboratory screening has identified sources of resistance in both wheat and barley. Oat, maize and sorghum are considered to be non-hosts, and while they may offer some option for reducing populations by rotation, the disease is not completely controlled.

STEM NEMATODES

Distribution

Ditylenchus dipsaci is by far the most common and important species of stem nematode on cereals, being widespread throughout western and central Europe, the United States, Canada, Australia, Brazil, Argentina and North and South Africa, although it is of greatest economic importance in temperate zones. Economic damage is rarely associated with sandy soils; soils with a clay base are more likely to be associated with damage.

Another species, *D. radicicola,* is distributed throughout the Scandinavian countries, the United Kingdom, the Netherlands, Germany, Poland, the former Soviet Union, the United States and Canada. This nematode also occurs on many grasses of economic importance.

Biology

Ditylenchus dispaci is a migratory endoparasite that invades the foliage and the base of the stem of cereals, where it migrates through tissues and feeds on adjacent cells. Reproduction continues inside the plant almost all year-round but is minimal at low temperatures. When an infected plant dies, nematodes return to the soil and from there they infect neighbouring plants.

Typical symptoms of stem nematode attack include basal swellings, dwarfing and twisting of stalks and leaves, shortening of internodes and an abundance of axillary buds, producing an abnormal number of tillers to give the plant a bushy appearance. Heavily infected plants may die at the seedling stage resulting in bare patches in the field, while other attacked plants fail to produce flower spikes. The nematodes are highly motile in the soil and can cover a distance of 10 cm within two hours, hence their ability to spread from one plant to another is rapid.

There are a number of biological races or strains of *D. dipsaci,* which are morphologically indistinguishable but differ in host range. Kort stated that the rye strain is more common in Europe and that the oat strain is more common in the United Kingdom. Rye strains attack rye and oats as well as several other crops, including bean, corn, onion, tobacco and clover, and a number of weed species commonly associated with the growth of cereals in many countries.

The oat strain attacks oats, onion, pea, bean and several weed species but not rye. Wheat is also attacked by *D. dipsaci* in central and eastern Europe. The species *D. radicicola* invades root tips of plants to form local swellings, which are characteristically spiral-shaped and easily confused with the galled root symptoms caused by *M. naasi.*

Economic Importance

Economic damage by *D. dipsaci* depends on a combination of factors, such as host plant susceptibility, infection level of the soil, soil type and weather conditions. The longer the soil moisture content in the surface layer of soils is optimum for nematode activity, the greater the chance of a heavy attack. This is a problem with cereal crops growing on heavy soils in high-rainfall areas. The nematode is economically important on rye and oats, but not on wheat and barley. Although few studies have examined the economic importance of this nematode, work on oats in the United Kingdom attributed a 37 percent yield loss to *D. dipsaci*.

Little is known about the economic importance of *D. radicicola*; however, under field conditions in Scandinavia it caused poor growth of barley and is known locally as *krok*. S'Jacob suggested that biological races of this species occur.

Control

The occurrence of different biological races or strains of *D. dipsaci* makes it a difficult nematode to control. The only economic and highly effective method is the use of host resistance, which has been summarised in table form by Rivoal and Cook. In the United Kingdom, the most successful oat crop has resistance derived from the landrace cultivar Grey Winter, which has also proven to be effective in Belgium. Rotational combinations of non-hosts, including barley and wheat, offer some control method for the rye and oat races of *D. dipsaci*. However, once susceptible oat crops have been damaged, rotations are largely ineffective.

OTHER NEMATODES

There are other plant-parasitic nematodes that have been found or are implicated potentially to cause yield loss on cereals, although their global distribution and economic importance to date has not been clearly defined.

FUTURE DIRECTIONS

There are several genera and species of nematodes that are of economic importance to small grain cereals. The current understanding of some nematodes, such as the cereal cyst nematode, *H. avenae*, is much more extensive than others with respect to both biology and control measures, mainly in the form of host resistance. Others, such as seed gall nematode, *A. tritici*, are relatively easily controlled with the adoption of seed hygiene. Unfortunately, knowledge is limited with respect to the basic biology and control options for most of the other important nematodes described.

In the future, the ability to reduce yield losses caused by nematodes will require a greater understanding of many basic questions about pathogen biology and the application of appropriate control measures. The use of chemicals is an unrealistic commercial option for most cereal growers, and to date many of the cultural methods fail to offer complete control.

As a consequence, it is inevitable that breeding for resistance and perhaps tolerance is the major strategy for long-term and environmentally sound control of these pathogens. As stressed in this subject, in order to accomplish this, a sufficient understanding of pathogen biology and plant reactions is necessary. To capitalise on this information, it is necessary to combine research efforts, particularly for some of the more complex nematodes with race and pathotype differences; hence there is a great need for global collaborative research programmes. Furthermore, the adoption of molecular tools to assist both in pathogen identification and plant breeding will become an integral part of future research developments and ultimate control of these important pathogens.

DIAGNOSIS OF NEMATODE DISEASES

Most plant-parasitic nematodes occur in soil around roots and are ectoparasitic, but many endoparasitic species are found abundantly in rhizosphere soil. Some plant-parasitic species are not important economically since they do not cause significant damage to plants. When they do cause noticeable damage they are considered pathogenic.

Soil-inhabiting ectoparasitic forms become important when their population inflates to reach the so-called economic threshold. For any control programme, whether chemical, biological, physical or cultural, accurate nematode identification is of the utmost importance. Even determining pathotype or race may become important in a particular group, e.g. Heteroderidae. It is also important to know the hosts of a species and how important it is pathologically to plant growth and yield.

Major nematode parasites are polyphagous and invade many plant hosts, multiply quickly to have several generations per year and have easy means for spread and dispersal. Examples are seen in many root parasites and some parasites of above-ground plant parts (*Aphelenchoides ritzemabosi, A. besseyi, Ditylenchus dipsaci*). Seed-borne pathogenic nematodes (*Anguina tritici, Aphelenchoides arachidis, Ditylenchus dipsaci*) have developed special adaptive mechanisms, such as the ability to withstand desiccation, for their survival and dispersal. Species like *Rhadinaphelenchus cocophilus, Bursaphelenchus xylophilus, Deladenus* spp., *Fergusobia* spp., use insects as vectors/hosts.

Host specificity is shown only by some groups of nematodes, e.g. Heteroderidae, Meloidogynidae, Pratylenchidae, Neotylenchidae. Such host-specific nematodes are supposed to have co-evolved along with their hosts. Co-evolution and zoo-geographical distribution of plant nematodes can throw some light on their origin, evolution and radiation. Besides their normal mode of transmission, parasites, pests and pathogens are consistently being moved about by people. Human activities and cargo traffic by land, sea and air have tremendously increased the chances of dispersal of plant nematodes. Exotic pests are constantly being introduced into clean areas.'

The nematodes are brought in with exotic plants, imported grain, vegetable and fruit produce and in soil attached to packing material, machinery, tyres of motor vehicles, tools and even shoes. Live third-stage juveniles, which is the dispersal stage of the pinewood nematode *Bursaphelenchus xylophilus,* were recovered in Finland from pine boards used in a packing case imported from Canada. Dispersal through wind, rain and floodwater respects no national boundaries. The potato cyst nematode (*Heterodera rostochiensis*) possibly originated in the Andean region of South America and was imported into the United Kingdom during the Irish famine of 1845/46 on potatoes brought for resistance against *Phytophthora infestans,* the causal agent of the Irish famine. From there it was distributed to mainland Europe, India and Australia.

Similarly, *Heterodera avenae* on wheat was distributed by people. Many temperate nematodes are flourishing in high altitudes in the tropics because they were transported there with planting material. Application of detection and identification techniques are basic to checking for plant nematode introduction, spread and distribution. Knowledge of hosts and geographical distribution of important species is essential. For example false root-knot nematode, *Nacobbus* spp. and *Monotrichodorus* spp., occur only in the New World, although the former may occur under glass in Europe.

Heterodera zeae was found on *Zea mays* and described in India. Later it was discovered in Maryland, United States, on an indigenous monocot by a stream. It was immediately put under quarantine and it was thought that it had come from India. Since it was found on an indigenous plant in the United States, it must be indigenous to that country. Very likely it originated in Mexico and co-evolved with maize and other monocots and was distributed with maize by humans.

SURVEYING, COLLECTING, FIXING AND MOUNTING NEMATODES

Surveying large areas for the presence or absence of plant nematodes is important but difficult. Systematic surveys are conducted using statistical designs to obtain a reliable estimate of nematode abundance. Fields and patches in a large area should be selected at random and sampling procedure should be based on standardised sampling techniques.

Nematode population assays should be made to determine both the qualitative and quantitative occurrences of important species and to relate their population levels to crop damage. Non-agricultural areas adjoining cultivated land and non-crop plants in and outside that area should also be surveyed to identify potential pest problems. This will help an understanding of the alternative hosts and aspects of the biology of nematode species.

Biological data including damage symptoms, life cycles, population trends and agroclimatic information on soil types, climatic conditions, cropping sequences and the times of sampling are also important in surveys because populations are dependent on them. This knowledge is of utmost importance when control measures are contemplated. Surveys should have backing from diagnostic and identification centres to obtain quick and reliable identifications. Fields and large areas for surveys must be selected at random and statistical sampling designs must be used to obtain quick and reliable data on nematode prevalence.

Periodic large-scale surveys of important species are essential to assess and monitor nematode populations. They should be repeated every two to four years to determine the relative abundance of species and new introductions, and there should be a system of coordinating various surveys to have a clear picture of the host and distribution of important species within an area. Coordination of surveys and pooling of data at an appropriate centre for analyses are important. About 250 to 300 ml of soil from around roots is collected in a polythene bag and data on host, locality, soil type, etc. are tagged. Care should be taken to ensure that soil samples are moist during transit.

About 100 g of roots including sound, partially attacked and well-attacked parts, and a similar quantity of shoots showing apparent damage, are collected and stored in polythene bags. Soil from roots of adjoining grasses and weeds should not be allowed to get mixed with the samples. Auger samplers are used to obtain soil cores for population studies at different depths. Care must be taken that soil is not mixed with soil from above or below the segment to be analysed. Samples should be processed soon after collection. They may be stored at about 4°C for later extraction.

Extraction methods must be directed towards isolating all stages of nematode development. Variously modified Baermann techniques and the Cobb sieving and gravity method are used for obtaining quick and reliable results. Their aims are to recover actively migrating nematodes, and their drawbacks include missing out sedentary nematodes.

Many nematodes lie dormant in the soil or plant tissue and need soaking in water for several hours before extraction. A simple bucket-sieving method for isolating nematodes from soil samples of about 300 ml, using a large bucket and a 45 to 53m n aperture sieve. The nematodes thus extracted are separated from debris and mineral particles by pouring the suspension on to a wire mesh covered with tissue paper and submerged in water in a trough.

The active nematodes pass through the filter paper into clean water and can be concentrated by decantation or centrifugation. Soil and plant material can be treated similarly to extract live, motile nematodes. The centrifugal sugar-flotation technique and elutriators are good for processing several samples and obtaining all stages of active or inactive nematodes. Nematodes may be killed by pouring hot water (85° to 95 °C) over them. In doing so, nematodes should be in the smallest possible quantity of water so that they are exposed to instant heat. They may be fixed in 3 to 5 percent formaldehyde solution and preserved in the same medium.

Nematode structures are best studied when examined in water, fixative or lactophenol. For making permanent mounts, the nematodes are transferred to glycerine and mounted in the same medium. Several methods are described in Siddiqi for making permanent nematode slides. Siddiqi described a quick method of transferring nematodes to glycerine by processing through warm lactophenol.

The nematodes are transferred to warm lactophenol held in a cavity slide or watch glass. They will first shrink but later stretch out to their normal shape within two to three minutes. If they do not do so and remain shrunk, then more heat should be applied until they are fully stretched. Then the slide can be left at room temperature overnight.

The slide is again warmed and a few drops of a mixture of 75 percent glycerine and 25 percent lactophenol are poured on to the nematodes held in thick, concentrated lactophenol. The slide is kept warm for five to ten minutes. Finally the nematodes are transferred to glycerine. Sufficient time should be allowed for the glycerine to penetrate the nematodes, which then can be mounted in pure dehydrated glycerine.

For mounting, nematodes already in glycerine are transferred to a small drop of glycerine placed on a clean glass slide and surrounded by glass fibre pieces for support. Three small lumps of paraffin wax (melting point 55° to 60°C) are put around but separated from the glycerine and a 19 mm diameter cover slip is placed over the wax lumps. The slide is then heated until the wax melts and fills the space between the cover slip and the slide. Finally the slide is sealed with glyceel.

DETECTING NEMATODE DISEASES

Detection and identification of nematodes is the first step in controlling them and checking their spread. This may result in economic savings of great magnitude. For example, *Radopholus similis* was recognised as the causal agent of the spreading decline of citrus in Florida in 1953. Since then, regulatory agencies in neighbouring states have imposed stringent measures to control its entry and spread.

A constant vigil has saved the citrus industry of other states of the United States from being devastated by this disease. *Heterodera glycine* of soybean (*Glycine max*), occurs in the Far East and United States. It has been detected and identified in Egypt. Unless strict quarantine measures are applied, the nematode will soon spread in the countries of the Near East. Nematode diseases usually go unnoticed as infected plants are rarely killed.

A particular crop may show only patches of damage where the nematode populations are much larger than in the rest of the field. At times, entire fields may be infested and overall health of the plants may appear poor. The yield from such infested plants is considerably reduced. Farmers usually blame unfavourable soil or climatic conditions, but if the nematode disease is controlled, for example by the application of nematicides, yield is dramatically increased. General above-ground symptoms of nematode infection include unthrifty reduced growth, severe stunting, chlorosis and wilting of leaves and reduced quality and quantity of produce. Root damage includes swelling and knotting; shrivelling, splitting and cracking; brown necrotic lesions; and stubby, broken or girded rootlets.

Root symptoms are produced by three kinds of reaction:

— A hyperplastic reaction results in hypertrophy and hyperplasia in which both cell number and cell size are increased. Simple hyperplastic galling of root tips is caused by the feeding of *Xiphinema, Hemicycliophora* and *Belonolaimus* spp. Feeding by *Meloidogyne* spp. results in severe hypertrophy and hyperplasia of cells; multinucleate giant cells are formed at the feeding site in the cortex, endodermis and pericycle.

— A hypoplastic reaction that inhibits growth in meristematic tissue and causes stunted and stubby roots is caused by the feeding of tylenchorhynchids and trichodorids, respectively.

— A necrotic reaction is caused by nematode feeding, their movement within the tissues and the presence of micro-organisms. Browning and shrivelling of feeder roots is caused by ectoparasitic nematodes. *Pratylenchus* spp. and *Radopholus* spp. cause extensive necrotic lesions in roots by their feeding and movement. The necrotic reaction is caused by enzyme secretion into plant tissue and tissue response, as well as by the accumulation of phenolic compounds in the infected area of the root. Secondary invasion by bacteria and fungi aggravates damage.

Diseases may be caused by individual nematode species, or by a combination of several species or nematodes interacting with other pathogens to produce disease complexes. The disease complexes produced by the interaction of nematodes with pathogenic bacteria or fungi are more damaging to plants than these pathogens acting alone.

Some nematodes are vectors of plant viruses. *Xiphinema index* was first shown to be able to transmit the fanleaf virus from diseased to healthy grape-vines. Now we know of several longidorids and trichodorids which transmit the soil-borne viruses of plants. Nematodes have evolved various types of association with plants. The parasitic ones may be ecto- or endoparasites; they may be migratory or sedentary parasites. Many are host-specific but most are polyphagous. Siddiqi gave a detailed analysis of the ecological adaptations of plant-parasitic nematodes.

Migratory Parasites of Above-ground Plant Parts

Aphelenchoides besseyi, the causal agent of white tip disease of rice, is a migratory ectoparasite and is seed-borne. Infested seeds, when soaked in water for 24 hours, yield motile nematodes. The nematode also attacks above-ground parts of strawberries causing summer dwarf or crimp disease.

The bud and leaf nematodes (*Aphelenchoides ritzemabosi* and *A. fragariae*) feed ectoparasitically on buds and parenchymatous cells of the leaf mesophyll. *A. ritzemabosi* parasitises leaves of chrysanthemum and other ornamentals.

The symptoms produced include spots or blotches on leaves, distorted tissue and stunted and deformed buds. Soaking the infested plant tissues in water releases the live nematodes. Red ring disease of coconut is caused by *Rhadinaphelenchus cocophilus* in the Caribbean and northern parts of South America. It is transmitted to new trees by the palm weevil, *Rhynchophorus palmarum*. *Bursaphelenchus lignicolus* causes a destructive disease of pine trees in Japan. It infests axial and radial resin canals and is pathogenic causing the death of pine trees. It is transmitted by the cerambycid beetle, *Monochasmus alternatus*. Stem and bulb nematode *Ditylenchus dipsaci* was first described on teasel heads in Europe. It attacks stems of pea, beans, alfalfa etc. and bulbs such as onion (causing bloat disease), daffodil and narcissus.

In the Sudan and the Syrian Arab Republic, the infested bean and onion fields exhibit patchy areas and larger populations occur in wet rather than in dry areas. Heavily attacked onion and narcissus bulbs show rings of decayed tissue when they are cut. When infested stem or bulb tissues are macerated in water, hundreds of nematodes in various stages of development are liberated. In a diseased bulb, stem or seed, heavily infested with *D. dipsaci*, the pre-adult juveniles aggregate in thousands and roll up to form what is known as nematode wool which enables them to survive desiccation and other unfavourable conditions.

Ditylenchus angustus feeds ectoparasitically on rice stems, leaves and inflorescences producing various types of malformation of tissue. It causes Ufra disease of rice in Bangladesh and is known to occur in Myanmar, India and Madagascar. *D. drepanocercus* infests leaves of an evergreen tree, *Evodia roxburghiana*, in India.

Sedentary Parasites of Above-ground Plant Parts

Members of the family Anguinidae (*Subanguina, Anguina, Cynipanguina, Nothanguina, Orrina, Pterotylenchus*) produce galls on above-ground parts of plants; only one species of anguinids, *Subanguina radicicola*, produces galls on roots. *Anguina tritici* and *A. agrostis* form seed galls on wheat and grasses, respectively. *A. tritici* causes ear cockle disease of wheat resulting in wheat grains becoming deformed, brown or black galls that transmit the nematodes when sown along with healthy grains. Tundu disease of wheat is caused by the interaction of *A. tritici* with *Corynebacterium tritici.*

Anguina funesta produces seed galls on rye grass in Australia and acts as a vector of the toxin-producing bacterium, *Corynebacterium rathayi.* About 1 000 bacteria-infested galls are lethal to a sheep. Leaf galls are produced by *Subanguina, Cynipanguina, Nothanguina, Orrina* and stem galls are produced by *Ditylenchus, Pterotylenchus* etc. *Orrina phyllobia* is a parasite of silverleaf nightshade (*Solanum elaeagnifolium*) and can be used as a biocontrol agent for this perennial weed that infests cotton fields in southwestern United States.

Fergusobia spp., of the family Neotylenchidae, produce aerial shoot galls on *Eucalyptus* spp. in Australia and *Syzygium cumini* in India. They show dual parasitism of trees and agromyzid flies (*Fergusonina* spp.). Teasing out galls in water releases all stages of nematode development.

Migratory Ectoparasites of Roots

Most plant-parasitic nematode species are ectoparasitic. All plant-parasitic members of Dorylaimida, Triplonchida and Criconematina are ectoparasitic. In Criconematina, the male has a degenerate stylet and oesophagus and cannot feed. Migratory ectoparasitic nematodes may have small stylets (Tylenchidae, *Merlinius*) to feed on root hairs and epidermal cells, or long stylets (*Hemicycliophora, Amplimerlinius*) to feed on deeper tissues. Long and strong stylet-bearing *Hoplolaimus galeatus* and *Belonolaimus* spp. are pests of lawn and turf grasses in Florida. They produce yellow patches in which the root system is considerably reduced in size as a result of the feeding of nematodes.

The ectoparasites, *Xiphinema* and *Tylenchorhynchus,* kill epidermal cells by feeding and cause discoloration and superficial necrosis. Stubby root and docking disorder in crops are caused by trichodorid nematodes. Species of *Xiphinema* and *Longidorus* are known to transmit soil-borne NEPO viruses of plants. *Xiphinema index* and *X. diversicaudatum* transmit fanleaf virus of grapevines and arabis mosaic virus, respectively.

Longidorus elongatus is a vector of raspberry ringspot virus and tobacco black ring virus. *Trichodorus* and *Paratrichodorus* transmit TOBRA viruses of plants. Plants showing virus disease symptoms should be sampled for these nematodes. The ectoparasitic nematodes are generally polyphagous and survive in soil without a host for many years. They are easily extracted from soil by Baermann funnel, sieving or flotation techniques.

Sedentary Ectoparasites of Underground Plant Parts

Sedentary obese females of the ectoparasitic genera *Tylenchulus, Sphaeronema,*

Rotylenchulus, etc. are detected only by examining the roots. The author's report of the finding of *Tylenchulus semipenetrans* and *Rotylenchulus reniformis* in India in 1961 was the result of collecting the motile stages of these species by a sieving method using 250-mesh sieves.

The immature female stage and the male of the latter species gave rise to the idea of creating a new genus for them since they were unique to Hoplolaimidae. The author had not seen the kidney-shaped mature female, but the timely warning of Dr M.W. Alien of California University prevented a synonym of *Rotylenchulus* being proposed, as authors of *Leiperotylenchus* and *Spyrotylenchus* had done before. For identification of these nematodes, therefore, all stages of development should be examined. *Tylenchulus semipenetrans* causes slow decline of citrus and occurs in all areas of citrus culture.

The female's anterior end lies deep in the root reaching the stele region where it feeds, while its body, most of which remains outside the root, swells. The root tissues respond to nematode feeding by producing nurse cells near the head of the nematode, thus avoiding destruction and decay.

The nematodes may aggregate and form colonies on roots and lay eggs in a protective gelatinous material. Juveniles of *T. semipenetrans* remain viable in soil for several years after the citrus trees are removed. Recovery of these juveniles from the rhizosphere soil of plants that might have grown there is therefore not a proof of such plants being the hosts of this nematode.

Migratory Endoparasites of Underground Plant Parts

Pratylenchus and *Radopholus* spp. live, feed and multiply within the root and underground stem tissue. *Pratylenchus* spp. cause necrotic lesions in the epidermis and cortical parenchyma that may extend to the endodermis and stele. The lesions eventually girdle the rootlets and interrupt the water and nutrient supply of infected plants. Often such lesions are invaded by rot-causing bacteria and fungi, completely destroying the rootlets. *Radopholus similis similis* is a serious pest of banana, coconut, tea and black pepper. It causes severe root rot that, in the case of banana, results in black head or toppling disease, and in black pepper the yellows disease. *R. similis citrophilus* causes the spreading decline of citrus in Florida. The diseased trees show wilting of leaves, dieback of twigs and general unthriftiness, low fruit yield and a poor root system in which the feeder roots are shortened and reduced to less than half the normal size and number. *Scutellonema bradys* infests yam tubers and even multiplies in stored yams causing severe damage. *Hirschmanniella oryzae, H. mucronata* and *H. spinicauda* parasitise rice roots in many tropical and subtropical countries.

The nematodes feed and move about intercellularly in air channels between the radial lamellae and the parenchyma of the roots. *Hirschmanniella miticausa* parasitises taro (*Colocasia esculenta*) causing mitimiti disease in the southern highlands of Papua New Guinea. *Aphelenchoides arachidis* infests groundnut pod shells and seed testa in northern Nigeria, is resistant to desiccation and is seed-borne.

Sedentary Endoparasites of Underground Plant Parts

The root-knot nematodes, *Meloidogyne* spp., are obligate endoparasites and are very damaging plant pests particularly in the tropics. They are easily detected due to the galls or knots which they incite, usually on the roots, but occasionally on stems. Pearly white females, egg masses attached to the posterior of females and second-stage juveniles are recovered when a gall is dissected in water. *Meloidogyne* species found in marine habitats, e.g. *M. mersa* on *Sonneratia alba* roots in Brunei Darussalam mangrove swamps, burst in ordinary tap water and should be dissected out from the galls in saline or sea water. Galls are also produced on tomato stems, potato tubers and ginger and banana rhizomes as well as roots.

Meloidogyne species identification is mainly based on the characters of second-stage juveniles and the perineal patterns of the adult female. Root galling is not enough evidence for the presence of a *Meloidogyne* species. Juveniles in soil should be the best indicator of this nematode.

Sometimes *Meloidogyne* and heteroderid juveniles are found in soil around roots of a crop plant or tree, but repeated searches fail to locate the females. It may be that the females occur on roots of some associated weed or grass and are not a parasite of the crop being surveyed, or that these juveniles are the survivors of the populations from a previous crop.

The false root-knot nematodes, *Nacobbus* spp., are similar to *Meloidogyne* spp. in inducing root galls, but are very different in morphology. Since they occur only in the Americas, there is no problem in confusing them with *Meloidogyne* spp. in the old world. Cyst nematodes are also endoparasitic, but the mature females protrude from the root surface as they feed and grow, and when their bodies transform into brown or black cysts they can be easily dislodged from the root.

Heterodera and *Globodera* cysts are lemon-shaped and rounded, respectively. Cysts remain viable in soil free of hosts for up to eight to ten years. The cysts are recovered from the soil by flotation techniques. Cyst nematodes induce syncytia in roots from which they derive nutrition throughout their development. Vascular bundles are damaged by this host reaction and the flow of nutrients and water is disrupted. Because of their host specificity, the introduction of cyst nematodes and their spread can be checked by applying strict quarantine measures. Crop rotation with non-hosts is a good control measure.

PHYLOGENETIC RELATIONSHIPS

Taxonomy is the science of identifying and naming animals and plants and assigning them to groups in a system of classification. The identification and naming of individual taxon (plural taxa) is called microtaxonomy, while placing them in a hierarchical system of classification is macrotaxonomy. Taxa include any category from subspecies to superfamily. Taxa higher than superfamily, e.g. order or class, are not covered by the *International Code of Zoological Nomenclature.* Taxa are usually diagnosed on the basis of phenotypic similarity/dissimilarity and not on shared common ancestry. Macrotaxonomy as a science of classification is based on the philosophy of organic evolution.

All animals and plants are continuously changing, as are all non-living things in the universe. Some changes are brought about in a short time and are observable, but others take thousands of years and are not seen or sensed. Macrotaxonomy uses various methodologies to understand the mechanism and degree of this long-term change. These methodologies utilise the characters/features of animals/plants to determine their primitive or derived nature by assessing their weighting and homology. Conclusions are then drawn about their relationships for building a hierarchical system of classification. Relationships are interpreted in terms of descent from a common ancestor. This is not an easy task since interpreting homology increases the amount of speculation.

Several approaches have been developed to establish the relationships of taxa and to classify them. However, in classification, many nematologists are not using these approaches at all, but are largely depending on their own concepts and experiences of the groups. It is said that identification work, which is different from phylogenetic work, should not necessarily depend on such approaches, but new species and genera, when proposed, should always be properly diagnosed and the criteria for their differentiation be discussed to show their relationships with existing taxa.

MORPHO-ANATOMICAL METHODS

Species identification utilises characters/features of many kinds including morpho-anatomical, physiological, ecological, ethological, embryological and cytogenetical. By far the most useful and widely used characters are the morpho-anatomical. A comparative study of morpho-anatomical characters/features in a range of closely related species should be made to evaluate their importance in identification. This would include the study of each character's stability/variability within that group.

Hasan states: "Many species occur in mixture in natural populations. In most cases, due to lack of data on interbreeding, reproductive biology, physiology, ecology and geographical distribution etc., the concept of *phenetic species* has been employed in nematode taxonomy for diagnosing a taxon. Such a concept has its own importance, however many of these species have been known to exhibit a wide range of intraspecific variations in morphology and dimensions which according to animal taxonomists is the function of geographical isolation, ecological/environmental stresses and genetic inequality."

The biospecies concept involves determining a species on the basis of whether populations interbreed if their members are morphologically similar to each other but different from other such populations. Difficulties arise when either the members of the populations are morphologically similar but do not interbreed, or they do interbreed but are morphologically dissimilar. *Ditylenchus dipsaci* populations from phlox and others do not interbreed successfully.

Are they then separate species? *Globodera rostochiensis* and *G. pallida* differ morphologically, but they may mate freely in laboratory experiments and are expected to do so in the field. A female could mate with several males and thus the offspring in a cyst may not represent a pure culture.

Interbreeding as a test for species recognition does not apply to parthenogenetic forms, and among plant-parasitic nematodes such forms are quite common. Morpho-anatomical determination of species is therefore of utmost importance in parthenogenetic forms and is the main method of identification. Its application to sibling species, however, becomes difficult. *Radopholus similis* and *R. citrophilus* are similar in morpho-anatomical details but not so in physiological and cytogenetic characters. In such sibling species, the concept of subspecies is worth while.

The two species may be called *Radopholus similis similis* and *R. similis citrophilus* to show their relatedness but to keep them as distinct forms. The occurrence of closely related species sympatrically which maintain their morphological identity gives them a good taxonomic status. Parthenogenetic species are prone to clone-formation because they lack bisexuality and hence the exchange of genetic material. Such clones differ in small morphological differences and thus naming them as separate species could be dangerous. Variations among large populations from different places should therefore be studied thoroughly. These variations are due to isolation, geographical separation, host effects and several physical and chemical factors. Species will, however, continue to be described on morpho-anatomical characters. The superspecies concept, although not used in plant nematology, is a good one and includes, under one name, morphologically closely related species, sibling or cryptic species and subspecies, e.g. *Globodera tabacum* can be a superspecies to include *G. tabacum, G. virginiae* and *G. solanacearum*.

MORPHOMETRIC AND MORPHO-ANATOMICAL CHARACTERS

Since plant nematodes are microscopic, morpho-anatomical characters have to be correctly observed and interpreted. Even taxonomists fail to observe some characters correctly. For example, some *Helicotylenchus* spp. show a distinct dorsal oesophageal gland but indistinct subventral glands and so they have been described under *Rotylenchulus/Orientylus*.

A solid-appearing anterior region of the stylet conus *of Tylenchorhynchus* and a conus having an asymmetrical lumen in *Histotylenchus* can only be studied properly when seen in lateral view. A post-anal intestinal sac was wrongly reported in *Tylenchorhynchus stegus* and *Quinisulcius himalayae* possibly because the granules in the lateral hypodermal chords looked similar to those in the intestine. Phasmids were reported in the tail region of the Tylenchidae and of *Ditylenchus* although they are absent in these groups.

Morpho-anatomical characters including their variability in many populations from many hosts and geographic regions provide the basis for creating new species. Taylor and Jenkins described *Pratylenchus hexincisus* sp.n. and *P. subpenetrans* sp. n. and Fortuner described *Aphelenchoides siddiqii* sp. n. because they studied a number of closely related populations. The same applies to the author's description of new species of *Monotrichodorus, viz. M. acuparvus, M. parvus* and *M. proporifer* from South America.

Morphometric data. Morphometric characters including the de Manian ratios should not be the sole basis for species differentiation because they are often very variable. Taylor and Jenkins found L, a, b, c, c′ ratios *in Pratylenchus* spp. as highly variable, but the V ratio showed a greater stability. The V ratio is a good taxonomic character which is used to differentiate species of many genera, e.g. *Pratylenchus, Paratylenchus, Rotylenchulus.* In several cases, overlapping ranges of many morphometric characters make it very difficult to use them for species differentiation.

Sex and sexual dimorphism. Presence/absence of males and of sexual dimorphism are used as differentiating features. Where males are absent, the female spermatheca is reduced and empty. Sexual dimorphism in body shape is a good character of Heteroderidae and Rotylenchulinae and in the anterior region (cephalic region, stylet and oesophagus) is a characteristic of the Radopholinae as against the Pratylenchinae.

Body size and shape. Body length, width and shape (cylindrical, tapering at ends, kidney-shaped, lemon-shaped etc.) and habitus at death (straight, arcuate, spiral etc.) serve as useful differentiating characters. It should be kept in mind that fixation and processing to glycerine may cause reductions, distortions and artefacts. Therefore, specimens in water or fixative should always be measured and studied.

Cuticle. Cuticle thickness, striation, annulation (both transverse and longitudinal), punctation and ornamentation, and cuticular modification (ridges, spines, scales, alae) serve as useful characters as do the lateral field and their incisures. Thickening of the cuticle at the tail tip is a good character for *Trophurus* and *Paratrophurus.* Longitudinal ridges characterise *Mulkorhynchus* while the presence of five or six lateral incisures are used to diagnose *Quinisulcius* and the Merliniinae, respectively.

Body pores are used as taxonomic characters in Trichodoridae and the excretory pore position is used as a taxonomic character. Amphids, deirids, phasmids and sensory papillae including male supplementary papillae are very useful in the taxonomy of all groups. Bursal ribs on the tail are diagnostic for the Aphelenchidae while the presence of hypoptygma around the cloacal aperture differentiates the Merliniinae from the Telotylenchinae.

Cephalic region. The shape and the degree of separation from the body of the cephalic region is used in species differentiation. Its transverse and longitudinal striation, the formation of a labial disc and the degree of sclerotisation of its framework are also important.

Stylet and oesophagus. By far the most diagnostic characters are found in the stylet and oesophagus. Length of the stylet and the relative length of its conus are useful, as are the size and shape of the stylet knobs. Siddiqi says that the importance of the oesophagus in the classification of the Tylenchida. Among the Tylenchina, the families Tylenchidae, Psilenchidae and Dolichodoridae have a basal bulb enclosing the three oesophageal glands, while in the Hoplolaimidae, Heteroderidae and Meloidogynidae the glands lie free in the body cavity extending over the intestine. In the Hoplolaimidae, the glands may extend mostly ventrally or dorsally, but in the Heteroderidae and the Meloidogynidae, they are always ventral to the intestine. The dorsal gland nucleus is larger than those of the subventrals in the Xiphinematinae, while it is smaller than the subventrals in the Longidorinae.

The presence of the orifice of the dorsal oesophageal gland within the median bulb is one of the diagnostic characters of the order Aphelenchida. In the same group, the presence of a distinct isthmus is the characteristic feature of the Aphelenchoidea. The shape and size of the cardia or oesophago-intestinal junction are useful in the Dorylaimida.

Intestine, prerectum, rectum and anus. The number of cells in a cross-section of intestine and the length of the prerectum are useful in the taxonomy of the Dorylaimida. The anus in members of the Aphelenchida is a large, posteriorly directed crescentic slit compared with the small and pore-like structure in members of the Tylenchida. Rectum length in respect to anal body width is also used as a character. A post-anal intestinal sac differentiates between *Bitylenchus* and *Tylenchorhynchus* and its length is useful in differentiating species.

Female reproductive system. Reproductive systems can be didelphic, monodelphic, pseudomonodelphic, prodelphic, opisthodelphic, monoprodelphic and monoopisthodelphic. The length of the reproductive branch as a percentage of the body length is important in some cases, but in others it is highly variable since it depends on the degree of development of the ovary. The shape and location of the vulva is a good character, as is the presence/absence of lateral vulval membranes and epiptygma. The shape and size of the spermatheca and its position in respect to the genital branch (axial or offset) are good diagnostic characters. When a branch is reduced to a sac, its length and the presence/absence of a reduced ovary serve as good characters, e.g. in *Pratylenchus*. The length of the mature ovary and the arrangement of oocytes are also important, e.g. in *Aphelenchoides*.

Male reproductive system. The size and shape of the spicules, gubernaculum and spermatozoa are useful characters. There is a distinct difference in the shapes of the spicula and gubernacula of the Tylenchorhynchinae versus the Merliniinae and the gubernacula of the Pratylenchinae versus Radopholinae. The shape of spermatozoa differs in species of *Radopholus*. Genital papillae and ventromedian supplements are important characters of trichodorid and dorylaimid nematodes. Copulatory muscles were used by Siddiqi to differentiate between *Trichodorus* and *Paratrichodorus* and the presence of hypoptygma was used by Siddiqi to differentiate *Merlinius* from *Tylenchorhynchus*.

Tail. Tail length and shape are used in taxonomy but they show considerable variation. The shape of the tail tip, although a useful character in many groups, was shown to be variable in some species *of Pratylenchus*. Nevertheless, a pattern of tail tip shape emerges when a large number of specimens are studied. For example, *Pratylenchus panamaensis, P. coffeae* and *P. loosi* have characteristic tail tip shapes. The tail is consistently elongated in the Tylenchidae, hooked in the Halenchidae and short and rounded in females of the Hoplolaiminae and males of the Heteroderidae and the Meloidogynidae.

Juvenile characters. Tail shapes in juveniles of various stages is a good character in the differentiation of *Xiphinema* species. Presence of spined juveniles in *Hemicriconemoides* differentiates it from *Hemicycliophora*. The shape and arrangement of spines in juveniles of this genus are used for species differentiation.

BIOLOGICAL, BIOCHEMICAL AND CYTOGENETIC METHODS

Biological characters, especially host preference, can be used in the identification of some plant-parasitic nematodes. However, the effects of physical, chemical and biological factors on host-parasite relationships are often great. Presence or absence of males is also a differentiating character, but it is known that at least in some species, males may arise as a response to environmental stress. Biochemical and cytogenetic techniques are used to discover the degree of genetic similarity in a taxonomic group. Such data must be obtained for several species to compare with both in-group and out-group members.

A comparative study of such data provides valuable information about characters/ features that are the result of common genetic material, as well as those that are unique for a particular member of the group. The uniqueness determines the identity of the taxon. Genetically dependent molecular data can be used concordantly with the morpho-anatomical characters to determine natural groupings and evolutionary trends. Isozyme phenotypes, particularly esterases, have been used in the identification of *Meloidogyne* spp.. An inventory of esterase/phenotypes is made for a comparative study. Although having great potential, polyclonal (PCA) and monoclonal (MCA) antibodies have been utilised so far only in a limited area for the detection and identification of nematode pests. Immunofluorescence and antisera could provide valuable tools for the detection of nematode diseases and identifying and classifying plant-parasitic nematodes. Biochemical characterisation makes identification simple and quick.

Serological techniques using PCA and MCA have been tried and some specific antigens from endoparasitic nematodes have been identified and used, for example in separating *Globodera pallida* from *G. rostochiensis*. Davies and Lander found a large number of MCA using three *Meloidogyne* spp., *M. incognita, M. javanica* and *M. arenaria,* but none was specific and each cross-reacted with at least one other species. However, three MCA were promising in distinguishing between the three species on ELISA-based and immunoblotting assays. Cytogenetic techniques to determine chromosome number have been used successfully in the classification of higher taxa and the identification of species and pathotypes of *Meloidogyne* and Pratylenchidae.

All the Criconematoidea have a basic chromosome number n=5. Dolichodoroidea have n=8 and Heteroderidae and Meloidogynidae have n=9 (polyploidy 18) and n=18 (polyploidy 36, 54), respectively. *Radopholus similis citrophilus* was differentiated from 7?. *similis similis* in having n=5 compared with n=4 in the latter. *Meloidogyne* species can be differentiated by determining chromosome number, differential host tests and studying perineal patterns.

Nematode identification

Nematode taxonomists

Experts are needed for the development and implementation of any scientific programme. Nematode taxonomists are few and far between and they have been branded as a dying breed. The importance of plant-parasitic nematodes in agriculture, the role of nematodes in ecosystems and the potential use and need for the preservation of biological diversity in non-agricultural lands, have increased the demand for nematode taxonomists. However, the demand cannot be met unless taxonomic teaching and training programmes are implemented and funds to support and sustain them are found.

Microscopes and computers

For any diagnostic work, a good compound microscope is essential. Fitting interference equipment to the stereoscope compound microscope provides sharp images and reduces eye strain. The use of the scanning electron microscope (SEM) has increased in the study of nematode species. New species descriptions often include SEM photomicrographs that show clearly such characters as the amphids, labial and cephalic papillae, lateral field incisures and genital papillae. However, for routine identification work, SEM is not essential. Diagnostic characters, if based only on SEM observations, will be difficult to use. Nevertheless, *Radopholus similis similis* and *R. similis citrophilus*, which are not differentiated by morpho-anatomical characters studied with a light microscope, are readily differentiated by examining the SEM photomicrographs of the cloacal region.

Computers are now being used for recording and analysing identification data. Survey records of parasites, hosts, occurrences, soil types, climatic conditions etc. become so large that manual sorting and analysis take an immense amount of time and labour. Also, it is difficult to edit and update the records. All such records and data can now be stored in computerised filing systems providing databases that can be searched, retrieved and sorted within a few minutes. It also avoids storage problems and the possibility of accidental damage. Databases can easily be copied into duplicate diskettes to avoid accidental loss.

REFERENCES

Atkinson, H.J. 1973: The Respiratory Physiology of the Marine Nematodes *Enoplus brevis* (Bastian) and *E. communis* (Bastian): I. The Influence of Oxygen Tension and Body Size. *J. Exp. Biol.* **59**(1): 255-266. PDF fulltext

BBC News 2003: Worms survived Columbia disaster. Version of 2003-MAY-01. Retrieved 2008-NOV-04.

Bell, G 1982: *The Masterpiece of Nature: The Evolution and Genetics of Sexuality*. University of California Press.

Goldstein, Jerome. 1998. "Compost suppresses disease in the lab and on the fields". *BioCycle*. November. p. 62–64.

Harrison, Una J., and Frank J. Louws. 1999. "Disease management through suppressive soils". *Department of Plant Pathology,* North Carolina State University (draft document). September 23. 14 p.

Hoitink, Harry A. 1986. "Basis for the control of soilborne plant pathogens with composts". *Annual Review of Phytopathology*. Volume 24. p. 93–114.

Hudson, Berman D. 1994. "Soil organic matter and available water capacity". *Journal of Soil and Water Conservation*. March–April. p. 189-194.

Ozores-Hampton, Monica, H. Bryan, and R. McMillian Jr. 1994. "Suppressing disease in field crops". *Bio-Cycle*. July. p. 60–61.

Sances, Frank V., and Elaine R. Ingham. 1997. "Conventional organic alternatives to methyl bromide on California strawberries". *Compost Science and Utilization*. Spring. p. 23–37.

6

Graft-transmissible Diseases

Budwood is the primary inoculum tissue used for most inoculations, but bark and leaves nay also be used. Budwood should not be collected during excessively hot weather ›ecause some CGTPs in the perimeter branches of field trees can be temporarily nactivated or severely suppressed by heat. When the season changes and temperatures ›ecome cooler, however, the pathogen will usually return from its reservoir location n the roots or shaded parts of inner branches.

An ice chest should be used for budwood storage when collecting. Clippers should ›e disinfected, when moving from tree to tree, by dipping or spraying in a 1 percent odium hypochlorite solution. Bark samples can be placed in a small plastic tube, but he tube should not be sealed. Immediately after collecting the tissue samples, they hould be put into polythene bags to prevent their drying and immediately put into n ice chest.

All samples should be labelled clearly at the time of collection. Upon arrival at the lant laboratory, they should be put directly into a refrigerator at 5-6°C. Avoid freezing he inoculum. Budwood can be maintained under refrigeration for two weeks or longer ut should preferably be used as soon as possible. If a field tree is selected as a primary andidate, a budstick should be taken below or proximal to a welldeveloped and typical ruit. A bud propagation is then made, and the propagation held in the greenhouse. his propagation will then become the primary plant, and budsticks can be taken nywhere from this plant for initial indexing, for heat-treatment, for shoot-tip grafting, r for use as positive control tissue to test the effectiveness of the heat-treated or shoot-p grafted plant.

NOCULATION METHODS

he most frequently used method for inoculating indicator plants for the detection of ıost CGTPs is by "bud"-graft inoculation. The term "bud"-graft includes buds with "eyes", stem pieces without "eyes" (sometimes called blind buds), and also chip-buds. The blade is first slashed through the inoculum tissue, and then a single slash is made in the stem of the receptor plant. This procedure is repeated ten to 25 times per plant. The slashed area of the receptor plant is then wrapped with budding tape. Citron is an excellent donor host as well as a receptor host for mechanical transmission by knife or razor-blade slash. In general, seedlings are preferred as receptor or indicator plants. However, if propagated clonal buds derived from seedling lines are substituted for seedlings, they should be tested and compared against the seedling for their performance as indicators since their performance as budlings may be different from that of seedlings.

Although "buds" are used as inoculum for most inoculations, other tissue and techniques, i.e. Ieaf, bark, root, or side grafts, should be continually tried and tested to find the most effective means of bringing out maximum symptom expression. This is especially true for any initial indexing of new diseases or diseases of unknown etiology. Specific clonal selections used as scion propagations rather than seedlings have been found superior as indicators for indexing of certain pathogens, i.e. the cachexia, exocortis and exocortis-like citrus viroids.

A vigorous rootstock such as rough or Volkamer lemon is recommended as a rootstock under the clonal bud. The forcing of clonal buds is recommended where tristeza is endemic and tristeza-susceptible indicators may show too strong a tristeza reaction, thereby masking symptoms of other pathogens. In many cases tristeza can be filtered out by inoculating trifoliate orange seedlings and using shoots of trifoliate as inoculum.

A modification of this technique is to graft an indicator scion bud on a trifoliate or citrange seedling, inoculating the seedling and forcing the indicator bud. In most cases tristeza will be filtered from the new growth of the developing indicator shoot. Some isolates of tristeza can pass through trifoliate or citrange, but most do not. When testing for the bud-union effect of citrus tristeza virus using a sweet orange scion budded on a sour orange rootstock, or for the bud-union crease of certain scions on trifoliate or citrange rootstock induced by the tatterleaf virus, propagation of the scion and inoculation of the rootstock can be done simultaneously and the sour orange or trifoliate rootstock seedling is then bent just above the scion bud to promote rapid forcing of that bud.

POSITIVE AND NEGATIVE CONTROLS

It is extremely important that both positive and negative controls be incorporated in each index test. A collection of infected source plants containing mild and severe CGTPs should be developed and maintained as a "virus bank". Sweet orange has been found to be an excellent holding or reservoir plant for almost all CGTPs. These reservoir or bank plants should be periodically indexed to ensure that the pathogen is present or has not changed. It is important that the mildest CGTP sources be collected and preserved in the "virus bank", and these should be used as positive controls for each index test. These positive controls will provide the determining factor as to when an index test should be terminated.

The inclusion of negative controls is also very important, and they should be generously incorporated into every index. Negative-control plants give an indication of possible environmental or insecticidal spray damage, and can show effects other than those induced by pathogens. They also act as a standard for plant-size comparison when subtle pathogens or diseases of unknown etiology may stunt index plants, but otherwise show no other leaf symptoms. However, their primary importance is to provide a normal control plant when reading for very mild leaf reactions in the inoculated plants. Thus, the presence of a new pathogen or a very mild form of a known pathogen can be detected. Although it may appear to be an extra use of seedlings, the presence of a non-inoculated control in each container has been found to be very helpful for a number of reasons, but specifically for judging any possible reaction in the inoculated plants in the same container.

TIME OF FIRST SYMPTOM APPEARANCE

The time in weeks from inoculation to appearance of the first symptoms under optimum growth and temperature conditions is given in the specific section for each disease. During critical flush periods, plants should be observed daily for development of symptoms for certain CGTPs. Symptoms of psorosis, psorosis-like pathogens and concave gum oakleaf patterns may disappear from the young developing leaves as the leaves mature, and symptoms may not reappear in later flushes. Leaf-flecking symptoms are best observed during the first to third flushes of growth. Plants should be watched carefully to catch the growth at maximum unfolding of the leaves for best reading of young leaf symptoms. Different pathogens will show leaf reaction at different times. Records should be carefully maintained for the time of appearance of symptoms with a description of the plant reaction.

Maintaining proper temperatures is extremely important for appearance of some symptoms. If the temperatures are kept too warm, certain "cool" temperature pathogens may not show symptoms in leaves, or show them very poorly. If temperatures are kept too cool, symptoms for diseases such as stubborn, cachexia, exocortis or certain citrus viroids, which require warm temperatures for best symptom expression, may not show or may develop poorly in indicator plants. Also, citron reservoir plants used for PAGE detection of citrus viroids may not build a high titre of viroid under cool conditions. These plants must be held at warm temperatures.

The liberal inclusion of mild- and severe positive controls gives a working indication of the proper time and temperature for symptom appearance. The lack of any symptom development in plants inoculated with these mild-positive controls would invalidate the index. Vigorous growth is important for production of good leaf and stem-pitting symptoms. Stem pits are poorly produced in poor unthrifty plants.

CHECKING INOCULUM SURVIVAL

Two to three weeks after inoculation, the wrapping tapes should be removed, the inoculum examined for survival, and the survival recorded. If tapes are cut with a knife or razor-blade, these tools should be disinfected in a 1 percent sodium hypochlorite solution between each cut. When buds are taken from mature wood of a dark-coloured budstick, or when bark inoculum is used, it is sometimes difficult to tell if the inoculum tissue is dead or alive.

A small slice or cut made into the brown bark surface of the inoculum will reveal the bright green colour of living tissue beneath, thus indicating that the inoculum is alive. If both inoculum "buds" are dead, the plant should be reinoculated, or new inoculations made to another plant. Generally, if one of the two inoculum "buds" is alive, the plant need not be reinoculated provided there are sufficient replications.

Records

A record sheet for each index must be kept. This should include: the experiment number, date of inoculation, source of the inoculum, indicator plants used, inoculum survival rate, reading dates, and a large space reserved for notes on observations. Records should preferably include temperatures and light conditions under which indexing was done, and any use of artificial lighting.

Indexing Using Field Trees

Certain indexes require a longer term for completion of the expression of the mildest symptoms. At such times the inoculated index plants growing in the plant laboratory (greenhouse) need to be set out in the field, or field trees need to be inoculated and observed. For example, in the long-term index for cachexia the mild-positive controls may show no symptoms in the greenhouse even after one year. Therefore, it is best to move the indicator plants to the field and plant them at close spacing until the mild controls show positive symptoms. Similarly, certain strains of exocortis or related citrus viroids may require a field test to show mild bark cracking on their trifoliate or Rangpur lime indicator rootstocks.

The testing of sweet orange on sour orange rootstocks for the classical quick-decline tristeza reaction may also require an extended period of time for typical tristeza decline symptoms to develop. The testing for cristacortis also requires long-term observation of plants or trees in a screenhouse or in the field. These indexes should be carried out in an environment where temperatures are conducive to best symptom expression. Again, mild- and severe-positive controls should be present. For certain diseases, trees in the field may have to be tested or inoculated to observe specific symptoms, i.e. testing for blight, or observing fruit for symptoms of impietratura.

TRISTEZA

Tristeza is possibly the most destructive disease of citrus. Many millions of trees on sour orange rootstock have been destroyed in Argentina, Brazil, California, Florida and Spain. The disease continues to spread into new areas, e.g. Israel and Venezuela, destroying citrus plantings where sour orange is the predominant rootstock.

There are many strains of tristeza, causing various field symptoms on different scions and rootstocks. Isolates selected from field trees may induce a reaction only in Mexican lime. Some will cause bud-union failure of certain scions on sour orange rootstock, and others can induce severe pitting or stunting and yellows in a variety of indexed seedlings. Tristeza stem pitting can severely injure scions of grapefruit and sweet orange directly, regardless of rootstock, by inducing a severe pitting and enlarged cheesy bark in the scion, resulting in smaller fruit, loss of production and debilitation of the tree. Tristeza stem pitting also seriously affects lime and may limit production in many areas.

Extensive slide and text reviews on many aspects of tristeza are given in *Description and illustration of virus and virus-like diseases of citrus*, edited by Bové and Vogel, and these are highly recommended for reviewing the tristeza and tristeza seedling-yellows diseases. The plant index is still invaluable for detection of CTV and its many isolates. Graft inoculation to indicator plants can detect tristeza when virus titre is very low.

The severe seedling-yellows and stem-pitting forms of tristeza can currently be distinguished from mild forms only in plants. Garnsey et al.proposed a standardized host-range analysis for evaluating the severity of tristeza isolates by rating decline, stem pitting and seedling yellows on different hosts. There are strains of tristeza that are difficult to detect in seedlings of Mexican lime but that can be easily detected by ELISA. A California tristeza isolate (T-519) is very difficult to identify in Mexican lime indicator plants grown under temperature regimes conducive to good symptom development, but it is readily detected by ELISA. This illustrates the value of, and need for, using more than one technique in a programme for the indexing of important foundation block trees.

Method 1: Field Diagnosis

The iodine test. If the rootstock is sour orange and the scion sweet orange, grapefruit or mandarin, a sudden quick decline and wilting, followed by defoliation, especially during the first warm weather of spring, would suggest possible infection by tristeza. A simple field test can be carried out to detect starch depletion in the roots or rootstock below the bud-union. The disappearance of starch from the roots is a result of girdling owing to killing of the phloem cells at the bud-union. Tristeza-induced starch depletion generally proceeds from the outer tips of the roots back toward the trunk of the tree. The application of iodine to the exposed cut surface of a root is a rapid method of testing for starch. The following method is taken from the papers of Bitancourt and Fawcett:

- Prepare a solution of potassium iodide and iodine by dissolving 1.5 g of potassium iodide plus 0.3 g of iodine in 100 cc of water. Keep in a coloured glass bottle, away from direct sunlight.
- Dig up roots (6-mm or 1/4-in diameter or smaller) from the outer margin of the tree and cut roots at an angle exposing the inner wood. Place a drop of iodine solution on the cut surface. Lack of development of a dark blue or black reaction suggests starch depletion, whereas a strong reaction indicates abundant starch.

A tree on sour orange rootstock showing quick decline or sudden dieback symptoms and with low starch in the roots would be suspect for tristeza. Verification should be made using ELISA or graft inoculation to Mexican lime seedlings, or both, or any of the diagnostic techniques mentioned in this handbook.

Inverse stem pitting in the sour orange rootstock. A section of bark removed from the bud-union area of a tristeza-infected tree will usually show inverse pitting on the inner surface of the bark with corresponding pegs on the outer surface of the exposed sour orange rootstock. This symptom is highly diagnostic for tristeza.

Stem pitting on scions. Pitting is associated with, and diagnostic for, tristeza. Pits may be seen in the trunk and branches, and vary from severe deep depressions to closely spaced and small. When pitting is very severe, small branches will snap off readily at the new growth joints. The outer bark may be cheesy and, when peeled, stems may show very small closely spaced or varying sized pits. These can be seen in grapefruit, grapefruit hybrids pummelo, tangelo, limes, various citrus hybrids, and sweet orange, but rarely in lemon, sour orange, trifoliate orange or mandarin. Severe pitting on grapefruit and sweet orange in the field is usually associated with chlorotic, tight, upright branch growth.

Method 2: Seedling Indexing

Seedling indexing to Mexican lime is still a very powerful tool for detection of tristeza virus. The small-fruited and somewhat seedy Mexican or West Indian lime has various common names such as kaghzi in India, baladi in Egypt, doc in Morocco, and gallego in Brazil. The name key lime is also commonly used. Seedlings of *C. excelsa,* citron, *C. macrophylla* or other citrus which show vein clearing and stem pitting can also be used as indicators. However, the Mexican lime is highly sensitive to tristeza and is the preferred indicator.

Collection of budwood. Collect budsticks from a minimum of four quadrants of each tree. For routine reindexing of important foundation or mother-block trees where there may be some danger of possible infection by vector transmission, collect from eight sectors of each tree.

Inoculum tissue. "Buds" (buds with eyes, blind buds or chip buds), leaf discs or leaf pieces can be used. Graft two inoculum "buds" or leaf pieces, or a minimum of five or six leaf discs per plant. Since CTV is phloem-limited, it is important that inoculum tissue contain phloem and that cut surfaces of phloem tissue of donor and receptor plants are in good contact.

Inoculation. Place inoculum "buds" or leaf pieces in the lower part of the test seedling, removing as few leaves as possible from the lower stems. The seedling can be cut back to 20-25 cm from the soil surface at the time of inoculation or at two to three weeks after inoculation when wrapping tapes are cut and the inoculum is observed for survival. The time to cut back is decided upon according to the specific environmental conditions in each plant laboratory. With the use of plastic wrapping tapes at the laboratory at Riverside, California, plants are usually cut back at the time of inoculation with very high survival rates of inoculum buds.

Indicator plants. As stated above, Mexican lime is the recommended general indicator for identification of all types of tristeza. To determine if the isolates will cause seedling yellows or stem pitting, inoculate grapefruit seedlings. These are highly sensitive to most CTV isolates, and are the preferred indicator in an initial index for stem pitting (SP-CTV) and/or seedling yellows (SY-CTV). Most seedy grapefruits can be used, and the Duncan variety has been found satisfactory as an indicator. If seedling yellows is found in the grapefruit indicator, then subinoculations can be made from the infected grapefruit to sour orange and sweet orange to determine the severity of the SY-CTV or SP-CTV isolate.

Sour orange is an excellent supplemental indicator for seedling-yellows tristeza and is equally as effective as the Eureka or Lisbon lemon. It is also preferred since it is highly polyembryonic and will produce 70 percent or more plantable seedlings compared with only 8 percent for the highly gametic lemons. Garnsey found that a clonal Eureka lemon grown as a cutting makes an excellent indicator for seedling yellows.

To determine whether an isolate will cause significant stem pitting in sweet orange, it is necessary to inoculate seedlings of a sensitive variety such as Madame Vinous. Although sour orange and lemon seedlings are excellent indicators for seedling yellows, they will rarely show stem-pitting symptoms.

Mexican lime used for general tristeza indexing and grapefruit and sour orange seedlings used for detection of severe CTV isolates can be grown three per container. Using two containers with three plants per container, two plants in each container are inoculated, leaving the third plant as a non-inoculated control. Sweet orange seedlings used for detection of severe CTV-SP isolates should be grown one per container.

Controls. The negative control plant in each container of three is not inoculated. A minimum of one mild- and one severe-reacting tristeza isolate should be included as positive controls in every test. The severe-reacting positive control isolate can be inoculated into two plants in one container and the mild-reacting positive control isolate inoculated into a minimum of four, but preferably six to eight plants. The mildestreacting isolate available should be used, i.e. one that induces very few leaf or stem-pitting symptoms in Mexican lime.

Positive controls for seedling yellows or stem pitting should be "buds" taken from a minimum of two reactive sources with known symptomatology and should be inoculated into two grapefruit and/or sour orange seedlings per container. A strong and a mild reactive isolate are preferred. In large-scale tests for seedling yellows, two or more known reactive isolates should be used. Negative controls must always be included.

Inoculum survival. The wrapping tapes should be cut and removed two to three weeks after inoculation and the survival of the graft inoculum recorded. Although tristeza is not readily transmitted mechanically, other pathogens are, thus it should be standard procedure to dip all tools prior to cutting into any plant.

Leaf-disc grafts can be evaluated one to two weeks after inoculation. If one of the two "bud" or leaf-piece grafts is alive, the plant need not be reinoculated. However, if both bud and leaf-piece grafts are dead, or if three or more of the five to six leaf-disc grafts are dead, then plants should be reinoculated or new plants inoculated.

Post-inoculation plant care. The developing young side shoots on the Mexican lime seedlings should not be trimmed for the first three flushes of growth, or for approximately eight weeks, in order to obtain the maximum number of leaves to examine for vein clearing.

Occasionally only a bottom shoot will have leaves with symptoms. However, after the third growth flush, the side shoots should be trimmed and the most vigorous terminal shoot tied to a stake and trained to grow as a single shoot for later examination for stem pitting. If not trimmed, Mexican lime seedlings have a tendency to send out many lateral shoots and, when grown three to a container, can quickly overcrowd the container and bench.

Single shoots permit more light, thicker stem growth and better pesticide spray coverage. Trimmed plants are also easier to handle, and a single vigorous shoot produces a large single stem that can be readily peeled for critical observation of stem pitting. Grapefruit, sour orange and sweet orange index seedlings used for seedling-yellows index tests should be trained as a single leader or shoot, starting with the first dominant emerging shoot. Seedlings of these varieties have a natural tendency to develop as single shoots, and little trimming is needed. Supplemental lighting during the winter will significantly aid the growth of lime and sour orange seedlings but may not benefit grapefruit seedling growth much.

Temperature requirements. Cool temperatures are necessary for maximum tristeza symptom expression in plants. Warm temperatures above 35°C may suppress development of vein-clearing and stem-pitting symptoms in Mexican lime seedlings. The preferred greenhouse temperatures for all tristeza and seedling-yellows indexing are 24-28°C maximum during the day and 17-21 °C minimum at night. Temperature control is especially important when checking for mild isolates.

Time of symptom development. Plants were inoculated with a number of tristeza isolates over a seven-year period at Riverside, California. Within the first five weeks, 88 percent of inoculated plants developed leaf symptoms, and within eight weeks 97 percent of plants were diagnosed as positive. Over 90 percent of seedlings developed seedling-yellows symptoms within nine weeks, and some showed symptoms after as little as five weeks. This is based on observation of 1200 positive control plants inoculated with seven selected severe isolates over a five-year period.

Symptoms of tristeza

Vein clearing. The primary symptom in both the young and mature leaves of the Mexican lime is intermittent translucent vein clearing. The symptoms are best observed if the leaf is held overhead so that sunlight shines directly through the brightly exposed leaf. For critical reading of mild leaf-flecking symptoms, plants may have to be taken out into direct sunlight if the greenhouse is shaded.

Vein-clearing symptoms in leaves of Mexican limes can be readily detected in plants inoculated with most CTV isolates. However, vein-clearing symptoms induced by some mild-reacting isolates may be difficult to see. Only a few leaves may show an occasional mild fleck in the vein. A non-inoculated control seedling in each container is most helpful in judging the vein-clearing symptoms in the leaves of inoculated plants. Mild vein-clearing symptoms do not usually persist in mature leaves. Plants should be observed frequently when new flushes are developing. The optimum period for observation is just as a leaf becomes fully expanded.

If the underside of leaves of lime or sweet orange is observed by reflected light, the veins will usually show distinct dark greenish-black, "water-soaked" areas. These same areas, when viewed through sunlight, will show strong vein clearing. These "water-soaked" areas may persist in mature leaves after vein clearing is masked.

Leaf-cupping symptoms are usually pronounced when plants are grown at cool temperatures and under good growth conditions, but they may not always be present. However, leaf cupping is also a symptom induced in leaves of Mexican lime seedlings by the vein-enation virus in the complete absence of tristeza and, therefore, in the absence of other symptoms, leaf cupping alone may not be diagnostic for tristeza. Leaf cupping usually remains after the leaf matures or hardens. It may be pronounced when plants are infected with severe CTV isolates.

Vein corking. Very severe isolates of seedling yellows tristeza may induce a corking on the veins of Mexican limes, sweet orange or grapefruit very similar to symptoms induced by boron deficiency.

Stem pitting. For most CTV isolates, pitting can be observed after about eight weeks by peeling back the bark and observing the peeled stem. However, pitting is best evaluated at the end of the index test about four to six months after inoculation when the bark of the large single shoots can be completely peeled and the stems carefully observed. If the bark does not peel readily, steaming the stems in an autoclave is helpful for loosening tight bark.

Most tristeza isolates, except perhaps the very mildest, induce pitting in Mexican lime seedlings. Pits may be few or numerous. The pitting symptom in the stems of Mexican lime or any other seedling indicator is highly diagnostic for tristeza. Observations of many thousands of tristeza-infected Mexican lime seedlings have indicated that stem pitting is strongly associated with vein clearing in the leaves.

Observation of stem pitting is very useful to confirm diagnosis when conditions are not optimum for leaf symptoms.

Symptoms of seedling-yellows tristeza. When three plants of grapefruit, sour orange or sour lemon are grown in an individual container and two are infected with seedling yellows. Stunting may be severe, moderate or mild. The leaves are usually smaller, chlorotic, sometimes yellow, and may have pointed tips, and the shoots are compressed, thus giving the plant a stunted appearance.

Symptoms of stem-pitting tristeza. Almost any seedling of any variety can be pitted by some specific tristeza isolate. Most CTV isolates, however, do not cause pitting in mandarin, sour orange, sour lemon, rough lemon or sweet orange seedlings or trees. Stem pitting can be very severe in limes, Citrus macrophylla, grapefruit, grapefruit hybrids, tangelos and certain pummelo cultivars, and may limit their use where severe isolates are endemic. Certain CTV isolates may severely pit sweet orange. The Pera orange of Brazil is particularly susceptible. and a severe stem-pitting isolate affects navel orange trees in Peru.

When grapefruit seedlings infected with seedling-yellows tristeza show severe stunting, pitting may be difficult to evaluate since infected seedlings are too small and stems too thin. Inoculations into larger seedlings may be necessary to force vigorous shoots if pitting is to be seen and judged. Some isolates of CTV will induce pitting without inducing stunting or yellows reaction in grapefruit or sweet orange. Occasionally severe CTV isolates may induce a thick "cheesy", bark, usually associated with many very tine pits. This symptom is highly diagnostic both in field trees and in inoculated plants.

Termination

Tristeza reaction in Mexican an lime. When shoots reach about 1 m (approximately four to six months from inoculation), one or two of the mild-inoculated positive control plants should be harvested, the bark peeled and the peeled stems critically observed for pits. If pitting is evident in the control plants which had been inoculated with the very mildest-positive isolates, and if vein-clearing symptoms were observed and recorded in past readings in these mild-positive controls, then all plants can be harvested, peeled and examined for pits. Pitting can he recorded as none, mild, moderate or severe on a scale of 0 to 3 or 0 to 5. If no pitting is found in the mild-positive controls, the plants should be cut back, new shoots forced and the evaluation procedure repeated.

If no pits are seen in the mild-positive control plants, ELISA should be used to verify presence or absence of CTV. A modified procedure has been incorporated into the Citrus Clonal Protection Program at Riverside, California, which combines both the plant index and ELISA. "Buds" from budsticks collected from foundation block trees are first inoculated into Mexican lime seedlings.

After about eight weeks at optimum temperatures, the lime plants are trimmed to single shoots, with the exception of a side shoot which is allowed to develop. When this side shoot reaches a length of about 20 to 25 cm, it is harvested, and the bark peeled and processed for ELISA. In this manner, the tissue used for the ELISA has been grown under optimum temperature conditions in the greenhouse, and the low titre problem associated with seasonal temperature variation is circumvented. This combined procedure of plant index and ELISA uses the best of both index tests to assure freedom from tristeza in prime budwood source plants.

The T-519 tristeza isolate, which showed little or no vein clearing or stem pitting, has consistently indexed strongly positive by ELISA. However, the seven isolates used in this study represent selected severe-reacting seedling-yellows. Milder-reacting isolates would take somewhat longer to induce reactions. Also, as mentioned before, some stem-pitting isolates may show no yellows reaction in grapefruit or sweet orange but may show severe pitting. Therefore, all grapefruit or sweet orange plants should be maintained until they reach about 1 m in height before harvest. When possible. known positive stem-pitting controls should be used.

The sour orange used for seedling-yellows indexing rarely shows stem pitting and need not be peeled in routine indexing. The index test can be terminated when the milder-reacting seedling yellows positive controls show definitive symptoms. This may not be until three to four months after inoculation.

Method 3: ELISA

The use of ELISA for detection and diagnosis of tristeza is now a well-tested and proven technique and should be incorporated into any indexing programme.It is an excellent technique for surveys and large-scale testing, for obtaining very rapid results and for verifying the presence or absence of CTV isolates that are mild reacting in indicator plants. However, at present it cannot be used to distinguish between various isolates of tristeza, nor should it be relied upon as the sole index for testing Important foundation or primary budwood sources. New methods of identifying CTV strains or isolates by use of differential hybridisation techniques may, in the future, provide rapid identification of certain seedling-yellows and stem-pitting isolates, which currently can be distinguished only by long-term plant index.

Method 4: Microscopic Detection of Inclusion Bodies

Inclusion bodies of CTV may be seen in sectioned tissue by light microscopy. This can permit a very rapid means of tristeza identification.

STUBBORN

Stubborn is found in most countries that grow citrus under desert or semi-arid conditions. It is destructive in the warmer areas of California and Arizona and in most countries of North Africa, the Near East and the Arabian peninsula. Stubborn has been reported in Turkey, Greece, Italy, Mexico, Spain? the Sudan and Pakistan. Stubborn disease is rare in cooler climates since both the vector and organism prefer hot temperatures. It is not found in warm subtropical areas, presumably because of lack of suitable vectors.

This disease was first noticed about 1915 in navel orange trees near Redlands, California' and named "stubborn" by E.R. Waite, a budder. J.C. Perry observed that buds refused to grow properly, and "some influence was transmitted to good buds that were used in topworking". The name "acorn disease" was also used to describe the disease because of the many acorn-shaped fruit on diseased trees. A similar disease called "little leaf' was reported in Palestine by Reichert, who illustrated the small shoot and leaf condition as well as the small and misshapen fruit. Fawcett et al.. first showed the transmissible nature of stubborn disease.

A mycoplasma-like organism in the sieve tubes of stubborn-infected citrus tissue was discovered independently by Igwegbe and Calavan in California and by Laflèche and Bové in France. Both groups of workers concluded that a mycoplasma, and not a virus, was probably the cause. Fudl-Allah, Calavan and Igwegbe in California and Saglio et al. in France were able to culture a mycoplasma-like organism in liquid and solid media. The organism was described and named Spiroplasma citri, thereby establishing a new genus of mollicute. Antisera have been prepared to cultured *S. citri* and used for detection in various assays, including ELISA.

It is a motile, helical mollicute with no cell wall and no peptidoglycan. The spiral or helical morphology and motility can be seen by phasecontrast or dark-field microscopy. For positive identification of the causal organism, the first test is to observe the motile spiral spiroplasmas in a drop taken from culture media and placed under a dark-field or phasecontrast microscope.

Transmission of *S. citri* in California is primarily by the beet leafhopper Circulifer tenellus, but also by Scaphytopius nitridus. *S. citri* was shown to be spread from weed or vegetable hosts to a wide variety of weeds or vegetables by leafhoppers. The weeds became infected, stunted and yellow, and when they dried up under warm or hot conditions the vectors containing *S. citri* moved from the weed hosts to citrus. Young citrus are more susceptible than older trees. Transmission is primarily from infected weeds to citrus and to a lesser degree from infected citrus to citrus. The leafhopper Neoaliturus (Circulifer) haematoceps appears to be the primary vector in Corsica and in certain countries of the Near East.

Method 1: Field Diagnosis

Stubborn-infected trees in the field appear compressed and stunted, sometimes severely so. Stunted trees remain small; they rarely recover or die. The fruit does not colour properly and the stylar end retains its green colour. The navel orange is the most susceptible to fruit greening. Stubborn infected fruit is usually small and distorted, and may also be acorn-shaped in appearance. Fruit of seedy varieties may have a number of considerably smaller, darker purple seeds or completely aborted seeds. The fruit may have an insipid taste. The foregoing signs in combination are diagnostic for stubborn. Grapefruit, sweet orange (especially the navel orange), tangelo, mandarin, lime and pummelo are affected. Trifoliate and trifoliate hybrids, lemons and limes appear more tolerant.

During a period of intensive indexing of stubborn-infected trees in the Coachella and Central Valleys of California, many hundreds of index tests were made from symptomatic trees in the field to indicator plants in the greenhouse. The correlation of field symptoms with positive transmission was extraordinarily high, and much of the later diagnosis was made just by diagnosing symptomatic field trees. However, where symptoms are found in new areas, either transmission tests to indicators or isolation and culturing of the organism should be carried out.

Method 2: Transmission to Indicator Plants

Inoculum tissue. The best tissues for transmitting the Spiroplasma from citrus to citrus are stem pieces 5-7 mm in diameter obtained from compressed and stunted branches, or from the very young leaves of a new flush of growth. Extensive indexing tests for stubborn and results of comparative tests indicate that best transmission is made with side grafts. Calavan et al. also showed that the stubborn organism is poorly distributed in symptomatic trees and best index results were obtained with tissue collected during the spring months. Ten budsticks, and/or six young shoots with small emerging leaves, growing from compressed symptomatic twigs are collected per test tree. Budwood and/ or young shoots are put in plastic bags and transferred immediately to an ice chest.

Inoculation

Side grafting. Briefly, two side grafts are put into each seedling; each piece of graft tissue consists of part of a branch approximately 4-5 mm thick and 3-5 cm long. A wedge cut is made at one end of the budstick, a cut made into the seedling, and the wedge fitted into the cut. The side grafts are then securely wrapped with polythene budding tape, and a sleeve cut from a polythene bag is placed over the area above and below the grafts to create a moist chamber.

Leaf grafting. A small rectangular section of leaf about 3 x 12 mm is cut from the midrib area of a young succulent leaf and placed into a T-cut in the bark of the seedling, as for standard bud-grafting. The area is then securely wrapped with polythene tape in the same manner as with buds. Two to three leaf grafts per plant are suggested.

Indicator plants. The Madame Vinous sweet orange seedling is recommended as a superior indicator for detection of stubborn disease. One seedling should be grown per container and trained as a single shoot to about 1 m, with a thickness of 5-7 mm. A minimum of five plants should be inoculated to index a given source tree; each plant may be inoculated with two side grafts, two to three leaf-piece grafts, or a side graft and two leaf-piece grafts.

Controls. It is helpful to have a known positive stubborn source plant growing in a warm room in the greenhouse. Stubborn-infected Madame Vinous sweet orange source plants have been held in the warm room at the Rubidoux laboratory (32-38°C maximum day temperature) in Riverside, California, for over 15 years, and have continually shown stubborn symptoms. Both stem and leaf parts taken from these plants have consistently transmitted stubborn over this period. Negative or self-inoculated control plants should be included in each index test.

Inoculum survival and post-inoculation care

Side grafts. Ten days to two weeks after inoculation, the bottom ends of the polythene sleeves are opened to permit partial drying around the side grafts. At three weeks, the polythene sleeves are removed and grafts observed for survival. The plants are then cut back to about 25 cm from the soil surface.

Leaf grafts. The wrapping tape surrounding the leaf graft is cut two to three weeks after inoculation, and inoculum survival rate recorded. Plants are then cut back and new growth trained to a single shoot as for side grafting. The leaf piece can be seen to grow within the T-cut of the grafted seedling.

Temperature requirements. Development of symptoms requires warm or hot greenhouse temperatures. Temperatures should be maintained at 32-38°C maximum during the day and not below 27°C at night. However, care must be exercised not to exceed 40°C for any length of time. This will result in development of small, abnormal leaves.

Symptoms. The first symptoms are a semi-wilted appearance of the young single shoot and leaves. The leaves are smaller, paler and slower-growing than those of the controls. The stubborn-infected shoot remains small and stunted whereas the new-growth leaves of the control shoot are much larger and upright. Leaves of stubborn-infected Madame Vinous indicators will develop translucent chlorotic areas near the leaf margins, especially in the vicinity of the tips, giving them a pointed appearance.

The chlorotic spots remain visible after the leaf matures. These leaf symptoms produced under warm greenhouse conditions are highly diagnostic for stubborn disease. As the plants continue to grow, negative control plants will develop normal vigorous shoots, but stubborn-infected plants grow very slowly and remain stunted and chlorotic, showing smaller leaves and closer internodes. Leaves may turn yellow or develop strong, typical zinc-deficiency-like spotted mottle.

Time for first symptoms and termination. Definitive symptoms can be expected eight to 12 weeks after inoculation. Under optimal conditions, symptoms may appear earlier. If no symptoms occur in the inoculated plants within 12 weeks, but the positive control plants show good stubborn disease symptoms, the experiment can be terminated. Stubborn-infected Madame Vinous plants generally show symptoms with the first emerging shoot and will rarely show a delayed positive response in subsequent growth flushes, provided plant growth and temperature conditions are optimal.

CACHEXIA

The citrus cachexia disease was named, described and first transmitted by Childs. Xyloporosis, a condition affecting sweet limes, has been linked synonymously with cachexia. However, cachexia is suggested as the preferred name, and xyloporosis is reserved for the specific condition or complex associated with sweet limes as originally described by Reichert and Perlberger.

The viroid nature of cachexia was first suggested by Roistacher et al., when many similarities between cachexia and exocortis were described, i.e. both pathogens: are highly mechanically transmissible and are readily inactivated on tools by sodium hypochlorite; cannot be eliminated from budwood by thermotherapy; are readily eliminated from microshoot tips by shoot-tip grafting in vitro; and react best in indicator hosts held under warm conditions. Cachexia is now known to be viroid-induced. It is a low-molecular-weight RNA consisting of about 300 nucleotides.

The cachexia disease is found in most citrus growing areas of the world where exocortis is found. Since most commercial citrus cultivars are symptomless carriers and the pathogen is readily transmitted in buds and highly transmitted mechanically by tools, the movement of this pathogen into new citrus-growing areas could present a problem, especially where mandarins, tangelos, tangors or C. macrophylla are grown as scions or rootstocks.

Method 1: Field Diagnosis

For photographs and a description of field symptoms of cachexia, Childs and Calavan and Christiansen. The diagnostic symptoms in tangelo, mandarin or *C. macrophylla* are phloem discoloration by gumming, undulating stem pitting or bumps and projections on the bark, which fit into depressions in the wood. Gum spots are usually prominent in the bark and are readily seen by slicing sections through the bark with a knife. The presence of these typical symptoms on susceptible hosts in the field can be diagnostic for the disease since very few graft-transmissible diseases of citrus show these classical symptoms, especially in tangelos, mandarins, Rangpur limes or *C. macrophylla.*

Method 2: Indexing Parson's Special mandarin or Orlando tangelo

Parson's Special mandarin forced as a scion on a vigorous rootstock is the preferred indicator for the detection of cachexia. The Parson's Special mandarin seedling grows too slowly to be an effective seedling indicator. However, if a bud from a seedling selection of Parson's Special mandarin is grafted on to a vigorous rootstock and forced under warm conditions, growth is vigorous and symptoms may appear at the bud-union interface between six and 12 months after inoculation.

The choice of a sensitive clonal seedling of the Parson's Special mandarin is very important. There is variability among seedlings and some are more sensitive than others. At the Rubidoux indexing facility, a number of Parson's Special mandarin seedlings were tested for sensitivity to various isolates of cachexia. Seedling selections Nos 9 and 10 were found to be the most sensitive and these are currently recommended. If the Parson's Special mandarin is obtained as seed, seedlings should be grown and observed for uniformity and vigour.

All slow-growing seedlings should be discarded and about ten of the larger and more uniform selections held. These should then be tested by forcing a bud from each of the selections as scions under a rootstock that has been inoculated with mild- and severe-positive cachexia, respectively. Alternatively it may be simpler to obtain selection Nos 9 or 10 as a clonal bud line. Inoculated seedlings of Orlando tangelo or pathogen-free lemons or grapefruit as scions budded to Orlando tangelo as the rootstock can be used as indicators for indexing in the field.

The Orlando tangelo as a seedling is preferred. The seedlings can be field-grown or preferably greenhouse-grown, "bud"-inoculated and removed to the field when convenient. The procedures for indexing using Parson's Special mandarin forced on rough lemon rootstock in a greenhouse, or using tangelo seedlings for field planting, are as follows:

- *Collection of budwood.* Four budsticks are collected from each of four quadrants of the field tree to be tested. Ensure that the collecting tool is disinfected in a 1 percent sodium hypochlorite solution between trees. Inoculum should be labelled, secured in a polythene bag and immediately placed in an ice chest. The budwood is later stored in a refrigerator at the plant laboratory.
- *Indicator plants.* A minimum of four but preferably six plants should be used to test each candidate or selection. "Buds" (buds, blind buds or chip buds) cut from the various collected budsticks should be distributed equally among the indicator plants. Budwood should be refrigerated for possible future use.

For the Parson's Special index, rough lemon seedlings (or other fast-growing, vigorous seedlings) are grown one seedling per container, as a single shoot to about 1 m height. The Parson's Special mandarin selection Nos 9 or 10 or a selected seedling line held as a reserve plant in the greenhouse is used as the scion budwood.

Controls. Mild- and severe-positive controls, plus non-inoculated or self-inoculated negative control plants, should be included for each test or experiment. A minimum of four plants should be inoculated for each control treatment. Six or eight plants should be inoculated with the mild positive control since this will determine when the test can be terminated.

Inoculation. The Parson's Special mandarin bud is grafted to the rough lemon about 20 cm above the soil surface and, when wrapping the bud, the "eye" of the bud should be exposed for forcing. A minimum of two inoculum "buds" are then grafted anywhere in the rough lemon rootstock below the Parson's Special scion bud. Inoculum "buds" should be completely wrapped. The seedling is then bent at a point just above the Parson's Special scion bud, and the top of the bent seedling can be tied to the base of the plant or placed under the container. This bending aids the rapid forcing of the scion bud. Again, knives should be disinfected in a 1 percent sodium hypochlorite solution before moving to a new source of inoculum.

Tangelo seedlings are graft-inoculated with a minimum of two inoculum "buds" and these are completely wrapped. Knives should be disinfected in a 1 percent sodium hypochlorite solution before going to a new source of inoculum.

Inoculum survival. After two to three weeks, the budding tapes are removed from the inoculum, usually by cutting with a knife or razor-blade, which must be disinfected between plants in a 1 percent sodium hypochlorite solution. Inoculum survival is recorded and plants with dead buds should be regrafted, preferably with fresh material or with inoculum stored in the refrigerator. Where Parson's Special mandarin is used as the scion bud, the wrapping tape may be left on until the Parson's bud is well forced.

Post-inoculation care

Parson's Special mandarin index. The Parson's Special scion is trained to grow as a single shoot or leader, and is staked and tied. After the scion reaches about 1 m, it is cut back about 10 cm above the bud-union and a single bud near the top of the cut-back area is then forced and trained to grow as a new single leader. This new growth is then staked, tied and grown to about 1 m high in the same manner as the previous shoot. This procedure of growth, cut back and regrowth is repeated until the test is complete or until plants are removed to the field for further observation.

Plants should not be crowded on the bench. Adequate light is needed to maximise growth and symptom development. Each plant should have about 400 cm² (20 x 20 cm) of growing room.

Seedling index. After survival of inoculum is verified, the seedlings should be cut back at about 25 cm from the soil surface. The new growth should be trained to a single shoot or leader by removing all newly developed side branches. The young inoculated indicator plants are held in the greenhouse until they are large enough for transplanting to the field (about 1 m tall). The field location should be as warm as possible for maximum symptom development (symptoms may develop poorly under cool temperature conditions). The young trees should be well watered, fertilised and given good care.

Plants in the field can be close planted, preferably at about 2 m apart.

Time for development of symptoms

Parson's Special mandarin index. Symptoms in plants inoculated with severe isolates of cachexia may appear six to nine months after inoculation. Those inoculated with mild isolates may take nine to 12 months or longer. If symptoms do not appear on plants inoculated with the mild isolates after 12 months, the plants should be moved to a warm field location and planted at intervals of 1-2 m. The young trees should be well watered and fertilised and given good care. Symptoms should be evident under the extended field index within one growing season (provided temperatures are warm or hot).

Seedling index to tangelo in the field. Symptoms may appear in eight to 18 months after field planting in seedlings inoculated with severe isolates. Those containing mild isolates may take two to six years, depending on the temperature and mildness of the isolate.

Temperature requirements. Plants under index for cachexia should be grown in as warm an environment as practical or possible. Recommended greenhouse temperatures are 32-40°C maximum by day and 27-30°C minimum at night. Symptoms induced by the cachexia viroid, similar to those induced by the citrus exocortis or other citroı viroids, are best expressed in indicators under warm growing conditions.

Detection of symptoms

Parson's Special mandarin index. After six to eight months, when plants are at maximum growth, the control plants inoculated with the mild- and severe-positive inoculum caı be examined for symptoms. This is done at two places on opposite sides of the bud union; the bark is lifted back and the area examined for the typical gumming symptom diagnostic for cachexia. Similar cuts can be made in the regrowth area surrounding th joint where the plant had been cut back. If no symptoms are evident at the bud-unioı or the first joint, the bark is replaced and securely rewrapped with budding tape. Plant can be re-examined in this manner every two to three months.

Seedling index. When the bark peels easily or is "slipping", the mild-positive contro plants are observed for symptoms. Ensure that knife blades are disinfected. Two or thre such windows can be made, preferably in the lower parts of the trunk to observ symptoms. If no symptoms are seen, the bark should be replaced and wrapped wit budding tape. This permits the peeled area to regraft and heal, leaving the trunk les damaged and available for more extensive bark cutting and peeling at future observatio times.

Termination. When 75 percent of the mild-positive controls show definitiv symptoms, the bark can be completely peeled from all of the test plants and examine for symptoms. The symptoms are then recorded and the test or experiment terminatec

TATTERLEAF

The tatterleaf disease of citrus, induced by the citrus tatterleaf virus (CTLV), was firs described by Wallace and Drake as a transmissible disease that induced mottled an tattered leaves in citrus excelsa indicator seedlings. Calavan, Christiansen and Roistache first showed the destructive potential of this disease to citrange rootstock when tatterlea infected tissue was graft-inoculated to satsuma mandarin budded on Troyer citrang rootstock. Meyer (Beijing) lemon trees, which were first imported into the United State from Beijing (China) in 1908, were later found to contain the tatterleaf virus. Man Meyer lemon trees worldwide that originated from the 1908 introduction probabl contain the virus, including many propagations and plantings of the original Beijin lemon in China.

The disease is endemic in mainland China and may be widespread. Tatterleaf diseas is widespread in Taiwan Province and in Japan and is probably present elsewhere whe Meyer lemon or other infected citrus have been imported from these countries. It ha been reported from South Africa in declining Shamouti orange trees on citrumelo rootstock. Most citrus species and all commercial cultivars are symptomless carriers of the virus. When the bud-union crease is severe, the tops may shear off at the union in high winds. Miyakawa and Tsuji report that some isolates of CTLV do not cause bud union crease. Trifoliate orange is immune to tatterleaf and the virus is unequally distributed in citrange.

Semancik and Weathers showed mechanical transmission of CTLV from citrus to cowpea and partially purified the virus. It was rod-shaped, 19 by 650 nm, and transmissible to 19 herbaceous hosts. Wallace and Drake suggested that two viruses were present since shoots of inoculated *C. excelsa* indicator seedlings would recover after showing symptoms, and these recovered shoots contained a transmissible agent which would react in citrange but not in *C. excelsa*. Recovered c excelsa shoots could be reinfected by the virus present in Meyer lemon buds and showed tatterleaf symptoms. They called the new virus "citrange stunt". Roistacher showed that recovered shoots of *C. excelsa* graft inoculated to symptomless carriers would eventually show the tatterleaf component. He suggested that the disease was one complex and the original name "tatterleaf' be retained to describe both diseases.

CTLV is difficult to eliminate from budwood by shoot-tip grafting but was eliminated from budwood by thermotherapy. Koizumi eliminated the virus from citrus tissue by combining shoot-tip grafting with thermotherapy. Recently, Navarro and coworkers succeeded in eliminating CTLV from citrus tissue by shoot-tip grafting in vitro.

The virus is readily mechanically transmitted from infected citron to citron by knife or razor cuts, and the virus can be inactivated on tools by dipping them in a 1 percent sodium hypochlorite solution. CTLV was noted as spreading from tree to tree at the South Coast Field Station in southern California, presumably by mechanical transmission. It is important that indexing for CTLV should be included in any programme for establishing primary foundation trees since many citrus species and commercial cultivars are symptomless carriers and the virus is highly mechanically transmissible. CTLV can be very destructive to citrus on trifoliate rootstock or their hybrids.

Method 1: Index to Indicator Plants

Collection of budwood. Collect a minimum of four budsticks from each quadrant of the tree to be indexed.

Inoculum tissue. "Buds" (buds, blind buds or chip buds).

Indicator plants. Seedlings of Rusk, Troyer or Carrizo citrange, citremon and C. excelsa are used. If tristeza is endemic, omit the C. excelsa because it is very sensitive to citrus tristeza virus, and symptoms of tatterleaf may be readily masked. Mexican lime in the absence of tristeza may show pronounced psorosis-like symptoms. Rusk citrange is more sensitive than Troyer or Carrizo citrange. Zhang et al. reported that Troyer citrange was superior to Carrizo citrange as an indicator. Indicators can be used as seedlings, as a clonal budline topworked or budded to a plant containing the virus, or budded as a scion on to any inoculated seedling and forced. Grow indicator plants preferably three per container and inoculate at least four plants, two in each container, leaving the third plant as a negative control. Where Rusk or other citrange buds are to be forced as scions, grow one rootstock seedling per container.

Inoculation. Use at least two "buds" to inoculate each plant. Cut back seedlings at about 20-25 cm above the soil surface. The seedling may be cut back at the time of inoculation, or two to three weeks later when inoculum survival is recorded. If citrange is used as a forced scion, the rootstock seedling should preferably be bent just above the scion bud or cut back or topped just above the scion bud.

Controls. Both positive and negative controls should be included. Where seedlings are used and grown three per container, one of the three should be left as a negative control for each index test. Positive CTLV control buds should be inoculated into two of the three plants in one container, leaving the third plant as a negative control.

Inoculum survival. Cut the grafting tapes two to three weeks after inoculation and record "bud" survival. Reinoculate only if both inoculum "buds" are dead. Since CTLV is highly mechanically transmissible on tools, dip the razor- or knife-blade used for cutting the grafting tape in a 1 percent solution of sodium hypochlorite between plants.

Post-inoculation care. After the initial cut-back, allow seedlings or the budded Rusk citrange scions to develop all shoots without trimming. Supplemental light will enhance growth of citrange during winter months and should be incorporated in the plant laboratory as normal procedure.

Temperature requirements. Tatterleaf is a cool temperature virus and warm temperatures have been observed to mask symptom development.

Maintain index room temperatures at 24-30°C maximum day (or preferably cooler), and 1821 °C minimum night.

Time for first symptoms. Five to seven weeks in C. excelsa and six to eight weeks or longer in Rusk or other citrange. Occasionally, delayed reaction may result in the appearance of symptoms beyond eight weeks.

Symptoms. The first symptom is a mild chlorotic spotting of the leaves, usually only in the first flush of growth. Later growth will show intense leaf spotting, leaf deformation and stunting. The leaves look "tattered" as if the edges were torn in a non-uniform pattern. Symptoms in citrange are clear spots that persist and develop into chlorotic spots. Stems are blotched. Symptoms persist in the mature leaves. Inoculated plants of *C. excelsa* may develop "recovered" or symptomless shoots after the first or second flush of growth. These are the symptomless shoots that will induce a reaction in citrange or citremon but not in *C. excelsa.*

Termination. Allow two to three growth flushes to develop, or wait until the positive control plants show strong and well-developed symptoms.

Method 2: Mechanical Transmission to Herbaceous Hosts

The method recommended for mechanical transmission is based on the procedure of Garnsey, which has consistently produced reliable index results. However, it must be remembered that sap inoculation techniques require higher concentrations of virus in the inoculum than is necessary for graft-transmission, and sensitivity is sacrificed for speed of assay.

Inoculation procedure. Young, succulent leaf tissue is macerated in cold neutral 0.05M potassium phosphate buffer at a ratio of 0.1 g of tissue to 1 ml of buffer. Plants are dusted with 500-mesh carborundum and leaves are inoculated with a cotton swab or fingers immersed in inoculum. Leaves are rinsed with tap water after inoculation and the plants are incubated in moderate light at 21-24°C.

Indicator plants. The cowpea Vigna unguiculata subsp. unguiculata (syn. Vigna sinensis) variety Early Ramshorn, other cowpea cultivars and the common red kidney bean can be used. Chenopodium quinoa has been used successfully in Taiwan Province. Grow five plants per container and inoculate four plants in each of two containers, with the fifth plant as the non-inoculated control. Bean and cowpea plants are usually ready to inoculate eight to 12 days after seed planting and under good growing conditions.

Symptoms. Symptoms will develop in four to six days. Symptoms in cowpea and red kidney bean are chocolate-brown necrotic lesions in the primary leaves. Mosaic patterns may appear in the secondary trifoliate leaves of cowpea under certain conditions. Symptoms in *C. quinoa* are chlorotic local lesions on inoculated leaves and a systemic mottle or mosaic. Citrus ringspot produces chlorotic local lesions in *C. quinoa* and is also systemic.

PSOROSIS COMPLEX: PSOROSIS-A, PSOROSIS-B AND RINGSPOT

The psorosis disease of citrus was first observed in Florida and California in the early

1890s and named psorosis based on the Greek psora = ulcer or mange. The disease was commonly called scaly bark. It originated in the Orient and was spread worldwide by the distribution of citrus species and varieties. Fawcett first proved transmissibility by graft-inoculation with buds from infected to non-infected trees.

Fawcett designated the disease as psorosis-A to distinguish it from a more virulent form called psorosis-B. Psorosis-A has been linked with the concave gum-blind pocket and crinkly-leaf infectious variegation diseases of citrus as one complex. However, recent evidence suggests they are all separate diseases, and their distinct differences in symptomatology and cross protection are in the respective sections. The psorosis disease was comprehensively reviewed by Timmer and Beñatena and reviewed and illustrated by Roistacher.

The agent responsible for the psorosis disease of citrus is almost certainly a virus, though it has not as yet been purified or characterised. Derrick et al.. reported that two components are associated with the citrus ringspot virus. These components can be isolated, and when mixed together are infectious. Cross-protection of the milder forms of psorosis-A against a challenge with the severe psorosis-B is diagnostic for classifying a disease as belonging to the psorosis-A complex. The major susceptible varieties showing bark scaling are sweet orange, mandarin and grapefruit. The sour orange, sour lemon and rough lemon show no external bark symptoms. Most citrus species and cultivars are symptomless carriers of the virus.

Citrus diseases that show bark lesions and also somewhat resemble psorosis-A are Rio Grande gummosis of grapefruit and leprosis of sweet orange. Indexing of budwood from suspect trees by graft transmission to specific indicator seedlings should readily distinguish these diseases from psorosis-A. The psorosis complex of viruses has many variants and contains isolates ranging from those which are non-mechanically transmissible to some which are, such as ringspot or the very serious and spreading Argentinian psorosis.

Twenty-one isolates of psorosis-A, obtained from various field trees at the University of California citrus variety collection at Riverside, were graft-inoculated into a variety of indicator plants and tested for mechanical transmissibility from citron to citron. There was much variability in symptoms found in citron, Dweet tangor and sweet orange among the various isolates, and only two isolates transmitted mechanically. However, despite this variability, all 21 isolates in sweet orange rejected a challenge from psorosis-B, confirming their identity as psorosis-A. The disease is spread primarily by humans via propagation of infected budwood. It has been observed that 50-year-old sweet orange tree showing no bark lesions in the field produced progeny trees with over 60 percent psorosis bark lesions. Thus, although psorosis may remain symptomless in certain host trees, the virus can be transmitted from a symptomless host and induce symptoms in progeny trees.

Mechanical, seed or root-graft transmission occurs with various psorosis types. Ringspot will move by mechanical transmission in the field. Currently a severe form of psorosis is spreading fairly rapidly in Argentina and is relatively destructive. The means of spread has not been determined, but vector transmission is suspected.

The first use of citrus seedlings to detect a graft-transmissible pathogen in citrus was by Wallace for detection of psorosis-A. This seedling index reduced the time required for indexing from an average of approximately 11 years, for development of bark lesions in field trees, to about six weeks for symptom development in the young leaves of sweet orange. This was a revolutionary development that pioneered this relatively rapid detection of citrus pathogens by indexing to greenhouse-grown plants.

There has been much confusion about the relationship between concave gum and psorosis diseases since both may cause leaf fleck in field trees or indicator plants. Roistacher and Calavan separate these two diseases for the following reasons:

— Psorosis-A causes a distinct bark scaling in sweet orange, mandarin or grapefruit, whereas concave gum causes concavities in sweet orange or mandarin trunks and branches and does not induce scaling. These are separate and distinct symptoms.

— A tree infected with the concave gum pathogen will usually show a series of concentric gum rings in a cross-section of the trunk or branch, whereas trees infected with psorosis-A usually show a specific wood staining in similar sections of the trunk or branch. These symptoms are very distinct.

— Many leaves on trees infected with the concave gum pathogen will show diagnostically strong oak-leaf patterns (OLP) in the developing leaves during the spring growth flush under cool conditions, whereas leaves on trees infected only with the psorosis-A pathogen rarely show symptoms in the spring growth flush except if co-infected with the concave gum pathogen. Leaves of trees infected with some of the more severe forms of psorosis, i.e. ringspot, may show strong patterns distinctly different from OLP in the young and mature leaves in the field, but rarely show typical OLP in the spring flush unless mixtures of pathogens are present.

— When inoculated into seedlings of sweet orange or mandarin, the concave gum pathogen induces a specific mild flecking in leaves of the early flushes, and the oak-leaf pattern usually develops in leaves of later flushes. Shock symptoms are rare. Inoculations with psorosis-A infected tissue will rarely show OLP unless in a mixture, and will usually show shock symptoms in the young flush of seedling growth under cool temperature conditions.

— Concave gum-infected sweet orange plants will not protect against a challenge from psorosis-A or psorosis-B infected tissue.

Method 1: Field Diagnosis

Psorosis-A can be diagnosed in the field if the two symptoms of bark scaling and wood staining of stems are observed. Bark scaling alone, though usually diagnostic, should not be totally relied upon for identification. Psorosis-like bark lesions, which apparently are not associated with a graft-transmissible pathogen, have been observed in several areas. The psorosis-B and ringspot forms of psorosis may show varying fruit symptoms or ringspot leaf patterns. These symptoms are highly diagnostic for the ringspot form of psorosis. Field trees showing bark lesion symptoms of psorosis-A in California, South Africa, the Mediterranean region, Brazil and occasionally elsewhere usually do not show leaf patterns on the young growth flushes on mature trees. Thus, field observation for leaf symptoms is not recommended for diagnosis of psorosis-A.

Method 2: Seedling Index

Indexing to seedlings is the principal method for the positive diagnosis for the psorosis-A complex.

Budwood collection. A minimum of four budsticks are collected from each quadrant of a candidate or test tree. If a tree in the field is selected as a prime candidate for thermotherapy or shoot-tip grafting, a budstick should be taken in the proximity of a typical fruit, and a bud propagation made from this budstick held in the greenhouse or plant laboratory. This greenhouse propagation becomes the primary plant, and budwood can be taken anywhere from this plant for indexing or for use as positive control tissue for testing the effectiveness of thermotherapy or shoot-tip grafting.

Indicator plants. The sweet orange seedling is the preferred indicator, and Pineapple, Madame Vinous or Olivelands sweet orange have been found to be superior varieties for detection of psorosis-A. There are definitive varietal differences in sensitivity to psorosis-A and such varieties as Koethen, Mediterranean or Diller sweet oranges should not be used. Mandarins are acceptable indicators but do not always show the shock symptoms associated with psorosis-A, and they may not be so sensitive to lesion formation in cross-protection tests. Although citron and lemon seedlings are excellent indicators for psorosis, they are sensitive to exocortis, other pathogens and tristeza, and psorosis symptoms may be masked or confused in the presence of these pathogens. If seedling-yellows tristeza is not endemic, the grapefruit seedling makes an excellent indicator for certain ringspot isolates.

At least four seedlings (two each in two containers) should be inoculated per test. These should be grown three seedlings per container; two are inoculated and one left as a non-inoculated negative control in each of the two containers.

Inoculum tissue and inoculation. A minimum of two inoculum "buds" (buds, blind buds or chip buds) per seedling are recommended. They are placed anywhere in the lower part of the seedling, leaving as many leaves in place as possible. The seedling can be cut off at about 2025 cm from the soil surface at the time of inoculation, or two to three weeks after inoculation when tapes are cut and bud survival rate is recorded.

Controls. As indicated above, a non-inoculated negative control should be included in each container. Also, if available, at least two psorosis isolates should be included as positive controls in each test: one mild and the other a known standard that consistently induces shock symptoms in inoculated seedlings.

Inoculum survival. Two to three weeks after inoculation, the wrapping tape should be cut and removed and the inoculum "buds" examined for survival. Although most psorosis isolates are not mechanically transmissible, some are, and mechanical transmission can be prevented by disinfecting the cutting tool by dipping it into a 1 percent sodium hypochlorite solution when going from plant to plant. Any dead or dying inoculum "buds" should be recorded, and if both "buds" are dead the plant should be reinoculated or a new plant used.

Post-inoculation care and temperatures. The temperature during the first four weeks after inoculation is critical for symptom expression. Cool temperatures will favour the appearance of shock reactions in the young emerging shoots, whereas warm temperatures may inhibit shock reactions and mask leaf symptoms. Psorosis-A symptoms are best expressed at relatively cool temperatures of 24-27°C maximum day and 18-21 °C minimum night. Shock and leaf-pattern symptoms may not appear if temperatures are too warm.

After the sweet orange seedling is cut back, the new growth should not be trimmed or suckered but permitted to grow or flush freely. The critical period for development of shock and young leaf symptoms is during the first and second flushes of growth. If possible, the use of insecticide spray should be avoided during this critical period of symptom development.

Supplemental light may enhance symptom development and should be used during the winter months.

Symptoms and time for development of first symptoms. Shock symptoms may show as soon as the first shoots emerge. Symptoms of shock are first observed in young emerging shoots as a wilt-like drooping followed by the drying up of these young "shocked" shoots. The leafless shoot is twisted, turns brown and necrotic and ultimately withers and dries up. It is later seen as a dried-up remnant which remains with the plant permanently. Shock can be complete, i.e. all emerging shoots dry up and lose their leaves. This is especially evident when plants are inoculated with severe isolates of psorosis-A or ringspot.

These appear four to six weeks after inoculation. As the leaves mature, the symptoms may disappear from the hardened leaves. Plants which lose all their first flush of leaves because of complete shock will usually show young leaf symptoms in the next flush of growth. The new emerging shoots of the second growth flush usually do not show shock symptoms. Shock symptoms may not occur if temperatures are too warm. Also the intensity of symptoms will vary depending on temperature and/or the isolate. Leaf symptoms may not occur or may be different if greenhouse temperatures are too warm.

Most of the leaf symptoms will appear in the first two flushes of growth. Later flushes may or may not be symptomatic depending upon the isolate and temperature. It is advisable to observe the developing leaves of the inoculated plants critically during the first two flushes of growth or approximately during the first eight weeks.

Termination. The presence of shock symptoms and young leaf symptoms in the positive controls. observed within the first eight weeks after inoculation, is sufficient to make a judgement of presence or absence of psorosis-A. If the object is only to ascertain the presence or absence of the psorosis pathogen, this eight-week period is sufficient, provided that the positive control plants inoculated with the non-heat-treated or non-shoot-tip grafted bud source are positive. If the object is to identify as fully as possible the reaction observed in the sweet orange test plants, then the three sweet orange seedlings growing in each container (two inoculated and one non-inoculated control) should be permitted to grow and be trained as single shoots. These inoculated plants will then be challenged with psorosis-B as given in Method 3.

Method 3: Cross-protection

To determine if the virus is psorosis-A, an inoculated sweet orange seedling is challenge inoculated with psorosis-B lesion inoculum and observed for evidence of cross-protection.

Challenging with inoculum of psorosis-B. A source of psorosis-B is needed for use in cross protection tests. This is obtained by grafting lesion inoculum, taken from trunk-bark lesions of a field tree, to a sweet orange seedling or budling. In six to eight weeks, under proper temperature conditions, blister-like lesions should form on the stems, and later develop on the leaves. The twig lesion inoculum is then used to challenge the sweet orange index plants which had shown leaf patterns, and to determine whether they belong to the psorosis-A family.

Challenge inoculation. Two "buds" from blister inoculum, are graft inoculated into one of the two pre-inoculated sweet orange indicator plants to be challenged, originating from Method 2. The non-inoculated control seedling is similarly inoculated, and the third plant is held as the pre-inoculated control. Plants can be cut back or left. Within eight to 12 weeks blister-like lesions should form on the challenged, non-protected control seedlings. The psorosis-B lesions usually form near the initial challenge inoculation site.

If lesions develop on the pre-inoculated and challenged test plants, then psorosis-A is not indicated. Conversely, if lesions do not develop on the pre-inoculated challenged test plants, but develop abundantly on the non-pre-inoculated challenged controls, then the virus in question is most probably related to the psorosis-A virus complex.

Method 4: Mechanical transmission of certain psorosis isolates and citrus ringspot virus

Citrus ringspot virus has been mechanically transmitted from citrus to citrus and from citrus to herbaceous hosts. The method given below for mechanical transmission from citrus to herbaceous hosts is based on Garnsey and Timmer.

Inoculum tissue and buffer. The best tissue is from young leaves of recently inoculated citrus plants in the shock phase. Leaves are ground in a pre-chilled mortar and pestle with cold 0.05 M TME tris buffer at pH 8.0, plus 0.5 percent (V/V) 2-mercaptoethanol.

Inoculation. Chenopodium quinoa is grown from seed in a well-fertilised and aerated soil mix to produce vigorous succulent growth. Artificial supplemental lighting should be used during the winter months. Almost fully expanded leaves of *C. quinoa* are dusted with 500-mesh carborundum and the inoculum applied with a cotton swab. Temperature. 21-27°C.

Symptoms. Symptoms will appear in four to six days as chlorotic local lesions distributed over the leaves.

Garnsey and Timmer succeeded in mechanically transmitting Florida, Texas and California ringspot isolates, plus three California psorosis-B isolates from citrus to Chenopodium quinoa. They could not mechanically transmit any isolate showing symptoms in C. quinoa back to citrus from *C. quinoa,* but could transmit ringspot isolates from *C. quinoa* to Gomphrena globosa and then back to citrus from *G. globosa.*

Mechanical transmission from citrus to citrus is best carried out by knife or razor slash into the stem. Citron is an excellent host and receptor plant. A slicing cut is first made into the branch of symptomatic citron tissue and then into the stem of the receptor citron. Ten slices can be used as a standard. The sliced stem is then wrapped securely with budding tape.

Many psorosis-A isolates will not transmit mechanically from infected sweet orange or citron to other citrus or herbaceous hosts. Therefore, a negative response to mechanical inoculations does not indicate freedom from virus, and does not eliminate the need for bud transmission to citrus indicator plants.

REFERENCES

Dick, J.B., and H.B. Tisdale. 1938." Fertilizers in relation to incidence of wilt as affecting a resistant and susceptible variety". *Phytopathology*. Volume 28. p. 666–667. (abstract).

Nelson, E.B., L.L. Burpee, and M.B. Lawton. 1994. "Biological control of turfgrass diseases". p. 409–427. In: A.R. Leslie (ed.). *Handbook of Integrated Pest Management for Turf and Ornamentals.* Lewis Publishers, Ann Arbor, MI.

Schneider, R.W. (ed.). 1982. *Suppressive Soils and Plant Disease.* The American Phytopathological Society. St. Paul, MN. 88 p.

7

Biocontrol of Soil-borne Plant Diseases

Biocontrol involves harnessing disease-suppressive microorganisms to improve plant health. Disease suppression by biocontrol agents is the sustained manifestation of interactions among the plant, the pathogen, the biocontrol agent, the microbial community on and around the plant, and the physical environment. Even in model laboratory systems, the study of biocontrol involves interactions among a minimum of three organisms.

Therefore, despite its potential in agricultural applications, biocontrol is one of the most poorly understood areas of plant-microbe interactions. The complexity of these systems has influenced the acceptance of biocontrol as a means of controlling plant diseases in two ways. First, practical results with biocontrol have been variable. Thus, despite some stunning successes with biocontrol agents in agriculture, there remains a general skepticism born of past failures.

Second, progress in understanding an entire system has been slow. Recently, however, substantial progress has been made in a number of biocontrol systems through the application of genetic and mathematical approaches that accommodate the complexity. Biocontrol of soilborne diseases is particularly complex because these diseases occur in the dynamic environment at the interface of root and soil known as the rhizosphere, which is defined as the region surrounding a root that is affected by it.

The rhizosphere is typified by rapid change, intense microbial activity, and high populations of bacteria compared with non-rhizosphere soil. Plants release metabolically active cells from their roots and deposit as much as 20% of the carbon allocated to roots in the rhizosphere, suggesting a highly evolved relationship between the plant and rhizosphere microorganisms. The rhizosphere is subject to dramatic changes on a short temporal scale-rain events and daytime drought can result in fluctuations in salt concentration, pH, osmotic potential, water potential, and soil particle structure. Over longer temporal scales, the rhizosphere can change due to root growth, interactions with other soil biota, and weathering processes. It is the dynamic nature of the rhizosphere that makes it an interesting setting for the interactions that lead to disease and biocontrol of disease.

The complexity of the root-soil interface must be accommodated in the study of biocontrol, which must involve whole organisms and ultimately entire communities, if we are to understand the essential interactions in soil in the field. The challenge in elucidating mechanisms of biocontrol is in reducing the complexity to address tractable scientific questions.

One of the most effective approaches toward the identification of critical variables in a complex system has been genetics. The study of mutants can be conducted in simplified laboratory systems or in the field, thus making accessible the examination of particular genetic changes and the associated biochemical characteristics in the real world. This subject presents recent advances in our understanding of the biocontrol of root diseases. We emphasize research aimed at enhancing our understanding of the biology of the interactions that result in disease suppression. It is this understanding that will make possible the practical use of microorganisms in the management of plant disease in agroecosystems. Numerous recent reviews present comprehensively the variety of microbial biocontrol agents.

The current and future directions in biocontrol, our goal is to present key themes in the discipline, drawing on the bacteria *Pseudomonas* and Bacillus and the fungi *Fichoderma* and *Gliocladium* as examples representing a range of life strategies and mechanisms of disease suppression. We address the principles of interactions of the biocontrol agent with the pathogen, the host plant, and the microbial community, illustrating each principle with some well-studied examples of successful biocontrol agents.

INTERACTIONS WITH THE PATHOGEN

Antibiosis

Biocontrol is often attributed to antibiosis. In many biocontrol systems that have been studied, one or more antibiotics have been shown to play a role in disease suppression. The fact that antibiosis is a common mechanism of biocontrol may be due to a bias in choice of organisms for study. Alternatively, it may be due to the attractiveness of the antibiosis hypothesis, or antibiosis may be simply a highly effective mechanism for suppressing pathogens in the rhizosphere.

Genetic analyses have been particularly informative in determining the role of antibiotics in biocontrol, in part because mutants can be screened easily in vitro for changes in antibiotic accumulation, providing the means to conduct thorough genetic analyses. Many antibiotics have been implicated in biocontrol only by correlative data; the following section focuses on those that have been implicated by mutant analyses and biochemical studies using purified antibiotics.

Role of Antibiotics

A number of highly effective disease-suppressive agents are found among the fluorescent pseudomonads, making this group of bacteria the most widely studied group of antibiotic producers in the rhizosphere. The first antibiotics clearly implicated in biocontrol by fluorescent pseudomonads were the phenazine derivatives that contribute to disease suppression by *Pseudomonas fluorescens* strain 2-79 and *P aureofaciens* strain 30-84, which control take-all of wheat.

Evidence for the role of phenazines includes an analysis of transposon insertion mutants that lack the ability to produce phenazine-l-carboxylate and are reduced in disease suppressiveness. Furthermore, the antibiotic is produced on roots grown in soil. P fluorescens strain CHAO produces hydrogen cyanide, 2,4-diacetylphloroglucinol, and pyoluteorin, which directly interfere with growth of various pathogens and contribute to disease suppression.

Mutants deficient in production of the antimicrobial substances are reduced in their ability to suppress certain diseases. Furthermore, a quantitative relationship between antibiotic production and disease suppressiveness is suggested by the enhancement of production of 2,4-diacetylphloroglucinol and pyoluteorin accomplished by adding extra copies of a 22-kb fragment of DNA that improves suppression of Pythium on cucumber.

The genes for the biosynthesis of many of the metabolites involved in disease suppression by fluorescent pseudomonads have been isolated, and their regulation has been studied. An emerging theme in the fluorescent pseudomonads is that global regulatory elements coordinate production of secondary metabolites. For example, biosynthesis of phenazine derivatives in P aureofaciens is regulated by a quorum sensor, PhzR, that perceives cell population density through the concentration of an autoinducer.

Interestingly, mutants that lack the ability to produce the autoinducer can use similar molecules produced by other rhizosphere inhabitants, suggesting that the presence of significant populations of other bacteria could influence phenazine production by P aureofaciens in the rhizosphere. Additionally, the environmental sensors ApdA and

GacA influence the production of numerous secondary metabolites involved in biocontrol by pseudomonads.

Sigma factors also regulate antibiotic production in fluorescent pseudomonads. Sigma factors are subunits of RNA polymerase that direct transcription in bacteria. Each sigma factor has promoter specificity and regulates a distinct set of genes, thereby playing a key role in gene regulation in bacteria. In *Escherichia coli*, gene expression in stationary phase involves sigma^S. A homolog of sigma^S is significant in biocontrol by *P* fluorescens through its role in expression of stationary phase genes required for the production of certain secondary metabolites and survival on plant material.

The ratio of the housekeeping sigma factor sigma and sigma appears critical in the regulation of various metabolites involved in disease suppression. The chemical, physical, and biological factors that regulate bacterial gene expression in the rhizosphere will contribute to the understanding of behaviour of biocontrol agents and will suggest strategies for improving their performance. Although bacilli have received less attention as potential bio-control agents than have the pseudomonads, evidence indicating that they may promote effective disease suppression is accumulating.

The bacilli are particularly attractive for practical use because they produce stable endospores, which can survive the heat and desiccation conditions that may be faced by biocontrol agents. One well-studied example is *Bacillus cereus* strain UW85, which suppresses diseases caused by the oomycetes, a group of protists that cause severe plant diseases. Analysis of mutants of *B. cereus* shows a significant quantitative relationship between disease suppressiveness and the production of two antibiotics, zwittermicin A and kanosamine.

Zwittermicin A is an aminopolyol representing a new class of antibiotic, and kanosamine is an aminoglycoside. The purified antibiotics suppress disease and inhibit development of oomycetes by stunting and deforming germ tubes of germinating cysts. Thorough analysis of antibiotic biosynthesis and regulation in the bacilli will depend on the development of genetic techniques, such as high frequency transformation, transposon mutagenesis, and reporter gene fusions, similar to those available for the pseudomonads.

Richoderma and Gliocladium are closely related fungal bio-control agents. Each produces antimicrobial compounds and suppresses disease by diverse mechanisms, including the production of the structurally complex antibiotics gliovirin and gliotoxin. Mutants of Gliocladium virens that do not produce gliotoxin are reduced in their ability to control Pythium damping-off. Mutants with increased or decreased antibiotic production show a corresponding effect on biocontrol.

Antibiotic Resistance

One of the goals of the use of biocontrol in agriculture is to avoid the pitfalls associated with synthetic pesticides, including the development of resistance in pest populations. An attractive feature of biocontrol strategies is that populations of pathogens resistant to antibiotics produced by biocontrol agents are likely to develop slowly. There are two reasons why this may be so. First, most biocontrol agents produce more than one antibiotic, and resistance to multiple antibiotics should occur only at a very low frequency.

Second, total exposure of the pathogen population to the antibiotics is low because, in general, the populations of biocontrol agents are localised on the root; therefore, selection pressures are minimised. Nevertheless, if sufficient selection pressure is applied to pathogens, the appearance of strains that are not controlled by biocontrol agents is inevitable. The use of antimicrobial agents in human medicine and agriculture has shown that selection pressures drive the evolution of resistance faster than we expect or hope for. The development of multiple-drug resistance derived from spontaneous mutations was not predicted to occur, but now it is recognised as a common mechanism of antibiotic resistance in bacteria.

Therefore, research is needed to understand the molecular bases of pathogen resistance to antibiotics produced by bio-control agents. Resistance should be studied before it occurs in the field so that fundamental knowledge can be applied to anticipate and prevent the breakdown of biocontrol. Examples of such approaches include inhibiting resistance proteins, combining antibiotics that select for different resistance genes, and avoiding use of biocontrol agents against pathogen populations in which a high frequency of resistance is predicted.

Similarly, understanding resistance of insects to *B. thuringiensis* has led to innovative strategies to reduce the impact of resistance on insect biocontrol. Resistance to antibiotics usually arises in a sensitive population by spontaneous mutation or by horizontal gene transfer. Mutations conferring resistance may affect antibiotic uptake or target sensitivity. Resistance genes that can be transferred encode antibiotic-modifying enzymes, resistant target molecules, or efflux pumps; such genes, when carried by antibiotic-producing bacteria to prevent them from committing suicide when they produce the antibiotic, are known as self-resistance genes. Self-resistance genes could be transferred to target pathogens in the soil; therefore, it is essential to understand mutations conferring resistance in target pathogens as well as self-resistance genes in producing organisms.

The importance of self-resistance is illustrated in the crown gall biocontrol system. Agrobacterium radiobacter effectively controls crown gall, which is caused by *A.*

tumefaciens, largely through the action of the antibiotic agrocin 84. Efficacy was threatened when agrocin 84-resistant strains of the pathogen were isolated from galls on A. radiobacter-treated plants. These strains were likely the result of transfer of plasmid pAgK84, which carries both agrocin 84 production and resistance genes, from the biocontrol strain to the pathogen.

A deletion in the plasmid-encoded functions for conjugal transfer generated a nonmobilisable derivative of pAgK84, which should provide more stable disease suppression. This example demonstrates both the possibility of antibiotic resistance in target pathogens and the importance of understanding the genetic basis for antibiotic resistance in the design of more robust strategies for biocontrol.

Despite the clear warning indicated by the transfer of agrocin-84 resistance, few other studies have focused on resistance to antibiotics involved in biocontrol. A gene encoding zwittermicin A resistance, zmaR, was cloned from the zwittermicin A-producing biocontrol agent €3. cereus UW85. The zmaR locus has been found in diverse *B. cereus* strains, including some that do not produce zwittermicin A, suggesting the possibility of horizontal transfer of zwittermi-cin A resistance within this species.

Although zmaR can confer zwittermicin A resistance on *E coli,* it is not known whether zmaR has been transferred to or can confer zwittermicin A resistance on microorganisms in the soil or rhizosphere.

Little information is available concerning spontaneous mutations that confer antibiotic resistance on pathogens that are the targets of biocontrol strategies. However, predictions can be made about the types of resistance mechanisms that might be deployed. For example, the wheat pathogen Septoria tritici acclimates to 1-hydroxyphenazine by inducing genes for catalase, superoxide dismutase, and melanin production; therefore, mutants of the pathogen that constitutively produce high levels of these protectants might not be suppressed by phenazine-producing biocontrol organisms.

The widespread occurrence of cyanide-resistant respiratory pathways in microorganisms suggests that prolonged application of hydrogen cyanide-producing biocontrol agents may select for pathogens containing cyanide-resistant oxidases. Recent evidence shows variation among strains of Gaeu-mannomyces graminis for sensitivity to antibiotics produced by fluorescent pseudomonad biocontrol agents, and disease induction by these resistant strains is not suppressed effectively by the biocontrol agent. It is not clear whether these resistant strains arose from a sensitive population by spontaneous rnutation or gene transfer or whether they are simply immune to the antibiotics because they lack an appropriate uptake system or sensitive target in the cell

Resistance due to mutations, gene transfer, or immunity will present a challenge for the use of biocontrol in the field.

Iron Competition

Biocontrol can involve suppression of the pathogen by depriving it of nutrients. The best understood example of this mechanism is iron competition iron is abundant in Earths crust, but most of it is found in the highly insoluble form of ferric hydroxide; thus, iron is only available to organisms at concentrations at or below 10^{-18} M in soil solutions at neutral pH. This presents a challenge for bacteria, which require iron at micromolar concentrations for growth. Bacteria have evolved high-affinity iron uptake systems to shuttle iron into the cell.

The typical system involves a siderophore, which is an iron-binding ligand, and an uptake protein, which transports the siderophore into the cell. The fluorescent pseudomonads produce a class of siderophores known as the pseudobactins, which are structurally complex iron-binding molecules. Analyses of mutants lacking the ability to produce siderophores suggest that they contribute to suppression of certain fungal and oomycete diseases.

An interesting aspect of siderophore biology is that diverse organisms can use the same type of siderophore. Microorganisms may use each other's siderophores if they contain the appropriate uptake protein, and plants can even acquire iron from certain pseudobactins. Further work is needed to characterise the ability of soilborne organisms to utilise side-rophores produced by biocontrol agents. Rapid breakdown of biocontrol would be expected if the target pathogens could circumvent disease suppression predicated on iron deprivation by acquiring the ability to utilise the siderophores from their neighbours in the soil.

Parasitism

In addition to antibiosis and iron deprivation, certain biocontrol agents also reduce plant disease by parasitising pathogens. *Trichoderma* spp parasitise fungal plant pathogens. The parasite extends hyphal branches toward the target host, coils around and attaches to it with appressorium-like bodies, and punctures its mycelium. These events require specific interactions between the parasite and fungal host, including the detection of chemical gradients and mycelial surface features. The specificity is illustrated by the observation that Trichoderma coils around *Pyfhium ulfimum* hyphae but not plastic threads of a similar diameter. Digestion of host cell walls is accomplished by a battery of excreted enzymes, including proteases, chitinases, and glucanases. These enzymes often have antifungal activity individually and are synergistic in mixtures or with antibiotics.

Mycoparasitism has been suggested as a mechanism of biocontrol by Trichoderma spp and *G. virens,* but its contribution to disease suppression remains uncertain, and clarification will be achieved with mutants lacking the cell wall-degrading enzymes. These mutants will be challenging to generate because Trichoderma harzianum produces at least three distinct chitinases as well as proteolytic and glucanolytic enzymes. Recent advances in molecular analysis, including the cloning of a 5: harzianum gene encoding endochitinase, and methods for transformation of Trichoderma and Gliocladium make the generation of mutants with multiple gene disruptions feasible.

Genetic Diversity among Biocontrol Agents

The complexity of the interactions involved in biocontrol and the wide range of environmental conditions found globally in agriculture make it unlikely that any one strain will suppress even a single disease in all settings. The genetic diversity of microorganisms with disease-suppressive potential remains a powerful yet largely untapped resource for biocontrol of plant disease. There is a need to seek new biocontrol strains, particularly strains adapted to the site where they will be used, but if random bacterial isolates from each site are screened for disease suppressiveness as has been done for years, the effort will continue to be labour intensive and will not make use of existing knowledge of biocontrol mechanisms. Recently, there has been some success in identifying diverse biocontrol strains that suppress plant pathogens through common mechanisms.

This approach has been employed successfully in insect biocontrol programs, most notably in the development of *B. fhuringiensis*-based strategies for the control of insects initial application of *B. thurin*-giensis produced limited and variable insect control. Biocontrol by *B. thuringiensis* is due to accumulation of a large protein toxin, which forms a crystal in the bacterial cell identification of the crystal toxin as the basis for insect control led to searches for genetically diverse *B. fhuringiensis* strains with related but unique crystal toxins. *B. thuringiensis*-based biocontrol strategies are now employed, and their development has generated a tremendous knowledge base of fundamental genetics and protein chemistry.

The limitation of this approach is that widespread use of organisms that share a mechanism of biocontrol will increase the selection for resistance; therefore, this approach must be coupled with an understanding of the mechanism and frequency of resistance and of strategies to avoid it. Recently, a comparable approach has been applied to bio-control of plant disease, and there have been some notable successes in finding diverse strains that suppress disease based on a common mechanism.

Genes for antibiotic production or antibiotic self-resistance are conserved among antibiotic producers and therefore form the basis of molecular probes for detecting new antibiotic-producing strains. Zwittermicin A-producing *B. cereus* strains and 2,4-diacetylphloroglucinol producing *P. fluorescens* strains have been identified from geographically and chemically diverse soils. Strains that produce the same antibiotic can be phenotypically diverse and may express traits useful under particular conditions. Some of these strains may provide effective control in certain soils in certain geographic regions or on particular crops.

In addition, the genetic diversity of these strains may be tapped by combining them in mixed inoculants. Certain mixtures of fluorescent pseudomonads, or fluorescent pseudomonads and fungi, suppressed disease more effectively than did single-strain inoculants.

INTERACTIONS WITH THE PLANT

Colonisation

It seems logical that a biocontrol agent should grow and persist, or "colonise," the surface of the plant it protects, and colonisation is widely believed to be essential for biocontrol. However, colonisation, or even the initial population size of the biocontrol agent, has been shown to be significantly correlated with dis-Base suppression in only a few instances.

In the suppression of damping-off of peas by *P. cepacia,* there is a significant relationship between population size of the biocontrol agent and the degree of disease suppression. Also, suppression of take-all of wheat is correlated with colonisation of roots by *P. fluorescens* strain 2-79. However, even in interactions that require colonisation for disease suppression, the biocontrol agent may not be required at high population density. It is intriguing that certain highly effective biocontrol agents, such as *P. cepacia* and *P. cereus,* achieve only modest populations on roots of field-grown plants and appear to replace, not augment, the population of indigenous members of their species. These results suggest that roots have a carrying capacity, or a limit on the size of a population they can support, for certain species of bacteria. Understanding the interplay between the sizes of the population of the biocontrol agent and the pathogen has proyressed significantly due largely to a theoretical modeling approach introduced by Johnson and augmented by empirical verification by others.

The application of mathematical models provides a depth of analysis that has not been achieved in past work on biocontrol. From this work, it is evident that careful attention must be paid to the dose-response relationship with both microbial partners, because biological outcomes are grossly different when observations are made at very high or very low disease pressure. Elaboration of these mathematical models will introduce a time variable, describing the relationship between disease severity or incidence and population of the biocontrol agent and pathogen during plant development.

The bacterial characteristics that contribute to active or passive spread on roots are not well understood. However, under field conditions, percolating water probably plays an essential role in the passive distribution of bacteria on roots. Nevertheless, motility is important for biocontrol in some plant-bacterial pairs under some but not all conditions, and osmotolerance is correlated with colonisation ability. Cell surface characteristics influence attachment to roots, which may be necessary for colonisation.

Certain mutations that affect accumulation of secondary metabolites also influence colonisation of plant material in field soil. A promising approach that will likely broaden the array of traits considered to be important for colonisation is to screen mutants directly for increased or decreased ability to colonise roots. Mutants of Pseudomonas strains of both phenotypes have been identified, and analysis of these mutants indicates that prototrophy for amino acids and vitamin B_1, rapid growth rate, utilisation of organic acids, and lipopolysaccharide properties contribute to colonisation ability.

The study of root colonisation by fungal biocontrol agents is more complex. Defining a fungal unit to quantify and the difficulty of conducting genetic analyses are challenges in studying the ecology of fungi. One study focused on mutants of Fichoderma that are resistant to the fungicide benomyl. These mutants are dramatically increased in root colonisation and biocontrol ability, even in the absence of benomyl. Benomyl resistance correlates with Several phenotypes, including increased cellulase production and altered morphology, making it difficult to determine the basis for increased colonisation. It is possible that increased cellulase production enhances root colonisation by enabling Fichoderma to utilise plant cell debris and that increased colonisation enhances biocontrol.

Induced Resistance

Some biocontrol agents induce a sustained change in the plant, increasing its tolerance to infection by a pathogen, a phenomenon known as induced resistance. In some cases, it is clear that induced resistance by biocontrol agents involves the same suite of genes and gene products involved in the well-documented plant response known as systemic acquired resistance (SAR), but this is not always the case. SAR is typically a response to a localised infection or an attenuated pathogen, which is manifested in subsequent resistance to a broad range of other pathogens. The best understood examples of induced resistance occur in the biocontrol of above ground diseases.

The idea that biocontrol agents might induce resistance in the host was first suggested on the basis of experiments showing that bacterial treatments protected potato tubers from subsequent infection by *P. solanacearum*. More recently, it has been shown that the bio-control agent *P. fluorescens* strain CHAO induces SAR-associated proteins, confers systemic resistance to a viral pathogen, and induces accumulation of salicylic acid, which plays a role in signal transduction in SAR. Mutants of CHAO that do not produce the siderophore pyoverdin do not induce SAR, suggesting a novel role for bacterial metabolites in disease suppression. Another fluorescent pseudomonad, *P. putida,* induces expression of the gene encoding PRla, which is associated with the classical SAR response. Other strains of *P. fluorescens* do not induce expression of the gene products associated with the classic SAR response but appear instead to induce a functionally analogous response.

Another line of evidence for induced resistance, which may or may not involve SAR, is that some biocontrol agents suppress disease when they are applied far from the site of infection by the pathogen, and they cannot be found at the infection site. Furthermore, in suppression of Fusarium wilt by *P. fluorescens,* preparations of lipopolysaccharides from the bacterial cell surface induce resistance as effectively as the living bacteria, demonstrating that biocontrol is not necessarily due to transport of the bacteria or an antibiotic through the plant. Whether or not biocontrol agents suppress disease by inducing resistance, it is essential that SAR and biocontrol strategies be compatible, because future agricultural practices are likely to require the integration of multiple pest control strategies.

Genetic Variation in the Host

Although much of the research focusing on plant genes affecting interactions with beneficial microorganisms deals with relationships with nitrogen-fixing symbionts, there is ample evidence that plants vary in their ability to support and respond to other beneficial microflora. The ability to support certain biocontrol organisms varies among plant species and among cultivars within species. Some plants appear to attract and support communities of microorganisms that are antagonistic to certain pathogens. Legume species vary in the magnitude of response to the plant growth-promoting bacterium *B. polynyxa,* and Bacillus strains isolated from wheat roots enhance growth of wheat in a cultivar-specific manner.

Plant species vary in their ability to induce genes for pyoluteorin biosynthesis in *P. fluorescens,* presumably due to variation in composition of root exudate among the species. Strains of *P. fluorescens* that overproduce pyoluteorin and 2,4-diacetyl-phloroglucinol provide superior disease suppression compared with the parent strain in some host-pathogen combinations and not others, and the effects correlate with host, and not pathogen, sensitivity to antibiotics.

Numerous studies have shown that different cultivars vary in survival or disease incidence in the presence of a pathogen and a biocontrol agent. The two challenges in assessing plant variation for this trait are to separate effects on the pathogen from those on the host and to partition host resistance and supportiveness of biocontrol.

The practical extension of the discovery that plants vary in the ability to support biocontrol is to enhance this characteristic through breeding. This has been referred to as breeding for "hospitality" of the host plant and is likely to have a substantial impact on efficacy of biocontrol of plant disease. Plants that are hospitable to a biocontrol agent might produce root exudates that support growth or induce expression of genes in the microorganism involved in disease suppression, attract the biocontrol agent to the infection site, or respond to the biocontrol agent by mounting a resistance response.

Breeding can also be employed to produce isogenic lines that can provide the basis for identifying traits in plants that influence their relationships with microorganisms. Breeding for hospitality to biocontrol agents will be facilitated by demonstrating heritability of the trait and by mapping genes associated with hospitality. This has been initiated in tomatoes, in which inbred lines derived from a wide cross have been assessed for their ability to support biocontrol by *B. cereus*. Substantial variation for the trait is observed among these lines, thus providing the basis for mapping genes that contribute to hospitality to *B,* cereus.

INTERACTIONS WITH THE MICROBIAL COMMUNITY

The interaction of the biocontrol agent with the microbial community may provide clues to explain why many organisms suppress disease effectively in the laboratory but fail to do so in the field. Biocontrol organisms may be affected by microbial communities, and they may influence the communities they enter, In some cases, they may enhance components of the community that work in concert to suppress disease. Recent evidence suggests that deliberate manipulation of microbial communities may be a highly robust and effective form of bio-control.

The ability of *P florescence* to suppress Fusarium wilt in radish seems partly due to its effects on the fungal community, in particular on the nonpathogenic strains of Fusarium oxysporum. Certain fluorescent pseudomonads have also been shown to displace resident fungi and bacteria, in some cases reducing populations of deleterious microorganisms. Treatments that enhance plant health can also increase the frequency of manganese-reducing bacteria in the rhizosphere community, thereby increasing manganese availability to the plant, which may in turn enhance resistance to disease.

Introduction of the biocontrol agent *B. cereus* strain UW85 can induce dramatic changes in the composition of culturable bacterial communities on soybean roots in the field. The change results in a community that more closely resembles a bacterial community from nonrhizosphere soil than does a nontreated root community. This finding, coupled with substantial support from previous work on the effects of host resistance and soil amendments on rhizosphere communities, suggested the "camouflage" hypothesis, which proposes that a mechanism for protecting plant roots from attack by pathogens is to make the roots "look more like soil by enhancing populations of typical soil inhabitants and reducing populations of typical root-associated microorganisms. The camouflage hypothesis has not been tested directly, but it may provide a useful basis for the study of community processes that affect biocontrol of root-associated pathogens.

There are three challenges associated with studying microbial changes resulting from the introduction of biocontrol agents. First, it is difficult to determine whether a community change plays a role in disease suppression or whether it is simply an unrelated outcome of altering the rhizosphere microflora. Second, the data generated in community analysis require new mathematical tools to deal with the complex communities and their multiple levels of interaction.

Caution must be used in interpreting results from purely descriptive studies. The power of such studies can be enhanced by application of multivariate statistical modeling. Finally, all of the work addressing the effects of biocontrol organisms on microbial communities has relied on culturing to describe communities. Because <1% of the bacteria in soil are culturable, studies based on culturing undoubtedly ignore some key organisms and interactions.

The recent application of molecular analyses to describe nonculturable communities in both extreme and familiar environments provides a powerful set of tools for the study of microbial interactions that influence biocontrol.

EFFECTS OF BIOLOGICAL CONTROL ON BIODIVERSITY

Biological control can potentially have positive and negative effects on biodiversity. Most of the time a biological control is introduced to an area to protect a native species from an invasive or exotic species that has moved into its area. The control is introduced to lessen the competition among native and invasive species. However, the introduced control does not always target only the intended species. It can also target native species.When introducing a biological control to a new area, a primary concern is the host- or prey-specificity of the control agent. Generalist feeders (control agents that are not restricted to a single species or a small range of species) often make poor biological control agents, and may become invasive species themselves. For this reason, potential biological control agents should be subject to extensive testing and quarantine before release into any new environment. If a species is introduced and attacks a native species, the biodiversity in that area can change dramatically. When one native species is removed from an area, it may have filled an essential niche, When this niche is absent it may directly affect the entire ecosystem. Because they tend to be generalist feeders, vertebrate animals seldom make good biological control agents, and many of the classic cases of "biocontrol gone awry" involve vertebrates.

For example, the cane toad (Bufo marinus) was introduced as a biological control and had significant negative impact on biodiversity. The cane toad was intentionally introduced to Australia to control the cane beetle. When introduced, the cane toad thrived very well and did not only feed on cane beetles but other insects as well. The cane toad soon spread very rapidly, thus taking over native habitat.

The introduction of the cane toad also brought foreign disease to native reptiles. This drastically reduced the population of native toads and frogs. "The cane toad also exudes and can squirt poison from the parotid glands on their shoulders when threatened or handled. This toxin contains a cocktail of chemicals that can kill animals that eat it. Freshwater crocodiles, goannas, tiger snakes, dingos and northern quolls have all died after eating cane toads, as have pet dogs ". This goes to show a small but deadly organism can alter the native biodiversity in an ecosystem in a very expedient manner. A pyramid effect can take place if native species are reduced or eradicated. The domino effect keeps on going and can potentially exude on other bordering ecosystems until an equilibrium is reached.

A second example of a biological control agent that subsequently crossed over to native species is the Rhinocyllus conicus. The seed feeding weevil was introduced to North America to control exotic thistles. However, the weevil did not target only the exotic thistles, it also targeted native thistles that are essential to various native insects. The native insects rely solely on native thistles and do not adapt to other plant species. Therefore, they cannot survive. Biological controls do not always have negative impacts on biodiversity.

Successful biological control reduces the density of the target species over several years, thus providing the potential for native species to re-establish. In addition, regeneration and reestablishment programs can aid to the recovery of native species. Native species can be affected in a positive way as well. To develop or find a biological control that exerts control only on the targeted species is a very lengthy process of research and experiments. In the late 1800's, the citrus industry was in great fear when the cottony cushion scale was discovered. This organism could cause a great deal of economic loss to the industry. However, a biological control was introduced. The vedalia beetle and a parasitoid fly were introduced to control the pest. Within a few years time, the cottony cushion scale was controlled by the natural enemies and the citrus industry suffered little financial loss.

Many exotic or invasive species can suppress the development of native species. The introduction of an effective biological control that reduces the population of the invasive species allows the rejuvenation of the native species. Biological controls can reduce competition for biotic and abiotic factors which can result in the re-establishment of the once over ran native species.

Effects on Invasive Species

Invasive species are closely associated with biological controls because the environment in which they are invasive most likely does not contain their natural enemies. If invasive species are not controlled, biodiversity may be at great threat in the affected area. An example of an invasive species is the alligator weed. This plant was introduced to the United States from South America. This aquatic weed spreads very rapidly and causes many problems in lakes and rivers. The weed takes root in shallow water causing major problems such as navigation, irrigation, and flood control. The alligator weed flea beetle and two other biological controls were released in Florida. Because of their success, Florida banned the use of herbicides to control alligator weed three years after the controls were introduced. Biological controls for invasive species also can have a negative impact on biodiversity.

The cane toad, as mentioned previously, is a great example of trying to control an invasive species. The cane toad was introduced to eradicate an invasive species. However, it became invasive, thus altering the biodiversity. The introduction of the cane toad could have potentially caused more of a disturbance in biodiversity than the targeted species did.

With further research and more scientific experiments, biological control could potentially play a huge role in the future of pest prevention. Biological control is being used among society today; however, it could someday reduce the use of many pesticides and herbicides. Since biological control could potentially have a large economic value, if found to be successful, research and job fields would increase continually. By increasing awareness of biological controls among more people, new successful biological controls could be discovered in the future. This could eliminate the overuse of chemicals. Biodiversity would increase, too, because of the reduction of chemical applications that often do affect not only the single species they are intended to kill, but other species as well.

Therefore, biological control is heavily analyzed by the amount of economic gain that directly comes from biological control. Many of the known economics of biological control are related directly to agriculture practices. Since agriculture has a huge impact on biodiversity this could potentially increase the biodiversity among agricultural practices. In order for agriculture to keep up with the growing population, many inputs are increased resulting in the loss of un-harmful species. Biological control use has been very minimal in agriculture. Less than 1% of global pest control sales of $30 billion involve biological methods.

A case study found that "calculated that releases of T. nubilale were considerably less cost-effective than pesticide applications used to control ECB on feed corn and fresh-market sweet corn. Pesticide applications produced 87% and 45% more net benefit (in dollars) than augmentation for feed corn and fresh market corn, respectively. In seed corn, however, Trichogramma releases produced essentially equivalent net benefits to pesticide treatments. In a third cost-benefit analysis of augmentation, Lundgren et al. showed that Trichogramma brassicae Bezdenko releases produced considerably less net benefit (94%; measured in cabbage head production) than methomyl treatments. In two other studies, "biological control releases were about two times the cost of pesticide applications; this was true for releases of a parasitoid, Choetospila elegans Westwood, used to control a stored product pest, Rhyzopertha dominica (F.) and releases of green lacewings, Chrysoperla carnea Stephens to control leafhoppers in grapes. Finally Prokrym et al.suggested that Trichogramma releases were about six times as expensive as pesticide treatments for O. nubilalis in sweet corn,". Another study shows that even though being possibly less effective, biological pest control still produces a benefit-to-cost ratio of 11:1. One study has estimated that a successful biocontrol program returns £32 in benefits for each £1 invested in developing and implementing the program, i.e. a 32:1 benefit-to-cost ratio.

The same study had shown that an average chemical pesticide program only returned profits in the ratio of 13:1. So while the exact numbers vary, the majority of these case studies shows that biological control is less cost effective than chemical applications and in result raises a flag that more research needs to be done. With progression in research, we can use more controls at a cheaper cost and increase the amount of biodiversity in areas because of the minimal use of chemicals that cannot target a specific species of pest.

REDUCTION OF INSECTICIDE APPLICATION

Forecasting potential outbreaks to avoid pest control operations has a long history in forestry. Once there was no insecticide problem because there were no insecticides. When they appeared forest entomologists like Escherich realized that the main purpose of pest control is to protect the crop and not to kill as many pests as possible. In his comprehensive books, very careful analyses of the complex biological situation surrounding outbreaks are presented. In some parts of Germany and in a few other European countries, supervision of pine forests based on the level of hibernating pests has been carried out for over a hundred years. The economic injury levels nowadays are known and critical density figures established for most forest types and main pests in Europe and North America. In preparing the forecast for potential outbreaks, not only quantitative data such as numbers of sawfly cocoons per unit area are to be considered. The qualitative aspects must also be properly taken into account, namely the effect of natural enemies must be measured and the trend of the gradation (outbreak) assessed. By using the technique of prognosis, which carefully weighs all available data of the complex and dynamic situation, many planned control operations have been cancelled or reduced when natural limitation factors were recognized early enough to be sufficiently effective. It is hoped that this type of preventive forecasting will be adopted in all countries having managed forests in order to minimize pesticide application.

Diagnosis of diseased specimens, rearing of parasites and hyperparasites, experience in the trends of gradations (outbreaks) and of epizootics as well as in the survival of partially defoliated trees - all these and many other factors are details in the whole picture of the situation. In other words, since foresters remained an ecologically-minded sector of plant producers, techniques are available to give reliable forecasts for the future development of incipient outbreaks. Thus, hasty or unnecessary control operations can be avoided.

The augmentation of natural mortality factors in the cultural and biological control of pest organisms can be achieved in forest management. Planned forest protection as part of general management is becoming widely acknowledged. To say it in a few simple words: the aim is to plant, to grow and to maintain forest types that will be, according to present experience, as pest-proof as possible within the framework of the economic task assigned to the forest.

The foundation of susceptible monocultures should be avoided and special care must be taken with the use of tree species and strains outside of their natural distribution area. Selection of more resistant strains of trees will be one possibility. Disruptive types of forest operations should he avoided. Sustained yield in forest production should become as important a target as productivity.

Silvicultural control is usually understood by foresters as any application of cultural measures to prevent existing pest organisms from reaching outbreak dimensions. The alteration of a susceptible forest into another, more resistant type can be achieved in the planning stage as well as later on, during maturation. Techniques of thinning, of protection against storm, sometimes even of fertilization can contribute to the general aim: to increase the environmental resistance of the forest to potential pest organisms. This type of control can, if successful, result in a low average level of pests and make other control operations unnecessary.

In natural populations of phytophagous arthropods, there are only two types of regulating processes known that are able to act more intensively as population density increases: intra-specific competition and natural enemy action. Under favourable environmental conditions for the pest, as with all plant-feeders, intra-specific competition cannot normally be utilized. A process that reduces populations because they have exhausted the available plant substance usually is of little practical control value in plant protection. Therefore, the only acceptable force, capable of some regulation, is natural enemy action; it has a key position in any integrated control programme. Actually, it would be difficult to imagine a situation in which integration of different control methods is planned without the direct or indirect role of natural enemies being carefully considered.

Acknowledging the central position of the effect of natural enemies, because they represent the only acceptable stabilizing and regulating element in an environment threatened by pest outbreaks, we immediately understand why all the methods of integration have two tendencies in common: either to preserve and augment natural enemy action, or to alter the environment so as to make it unsuitable for the pest. We understand also why the integration of natural enemy action with chemical control methods is so much more difficult than with any other control method - indeed, formerly they were considered to be totally incompatible. Recent developments have shown, however, that an optimum control pattern can be achieved in which the advantages of all available means, both natural and artificial, are integrated.

Several examples are known in which cultural or physical control measures have been specifically modified in order not only to spare beneficial agents, but also to rely on their help. Such old and direct methods as burning bark or sun-curing can be altered so that predominantly the pests and not their enemies will be destroyed. An example is the solar-heat treatment of felled ponderosa pine against the western pine beetle. Most cleric predators escape to the sheltered underside where they feed on the bark beetles that are not heat-killed. For this purpose, the bark must not be peeled from the exposed logs. Burning of the bark of trapping trees at times when the fewest parasites are present and leaving parts of the crown of felled trees in the forest to provide food for alternative hosts of important parasites are other examples.

To summarize these observations, many methods of pest control in the forest are known that do not rely on chemicals. Because of certain limitations of these methods and/or the very moderate scale on which they have been explored and tested, most integration problems arise through application of toxic chemicals in the forest.

In the development of a minimum insecticide programme it is desirable, but not yet frequently accomplished, first to lower the general level of the pest's abundance by utilizing more or less permanently acting natural enemies, assisted, when feasible, by silvicultural practices. When and where the level of pest abundance is still above the economic threshold, temporary rapid-acting biotic agents or chemicals could be applied in order to guarantee satisfactory control. There are several ways known to make insecticides less detrimental to natural enemies.

Insecticides can be physiologically selective, indicating an inherent difference of susceptibility between pest and enemy to a toxic substance; but very few are. Examples relevant to forest entomology are such outdated insecticides as chlorinated nitrocarbazoles. These are also stomach poisons and proved to be effective against such pests. Unfortunately, in spite of all suggestions by ecologists and wildlife protectionists, these compounds are no longer being produced.

The differential exposure of pest and natural enemies to the pesticide (physical or ecological selectivity) can be used to reduce the damage of broad spectrum insecticides and to achieve some degree of selectivity. It is here that modern integrated control efforts have revealed numerous new possibilities.

Several modern contact insecticides show some degree of selectivity that can be exploited. Endosulfan, for instance, is relatively harmless to bees, and systemics save the lives of non-plant-feeding arthropods, if properly applied to the tree trunk. Persistent pesticides are always more detrimental to the fauna, whereas reasonably fugitive residues allow the survival of entomophagous arthropods which enter the treated area from protected niches, as in the pupal stage in the soil, or immigrate from nearby untreated plots. Very little is known on the subject of pesticide degradation time with regard to natural enemies. These are usually physiologically more susceptible than the hosts and also more exposed to contact insecticides, due to their greater mobility.

Any minimum insecticide programme requires the application of chemicals only when and where natural limitation factors would fail with certainty ("spot treatment"). The philosophy behind this economy in ; the use of pesticides is not to reduce the cost of the treatment. It is frequently cheaper, per unit area, to treat large areas together. The difficult task of bioloigists participating in the planning of such operations—(and they should always participate!) is to convince—technically-minded foresters that the limiting of treatment to the most heavily attacked spots will be worth the extra effort and expense. As long as the treated areas are small, and the beneficial insects eliminated from the target area can immigrate from nearby refuges, the unbalanced situation will soon be buffered.

Minimizing the area treated, one of the established methods in integrated control, has no lower limit in forestry. It begins with the size reduction of treated patches in the forest, steps down to treatment of single trees, for instance those infested by bark beetles, and reaches an extremely dimension where persistent contact insecticides are applied, e.g. the tops of young pines, to protect only the buds of the leader against the bud moth, *Rhyacionia buoliana*; Miller and Neiswander, 1955. Another way to keep pine-needle-feeding caterpillars away from the crown is to put insecticide rings around the trunk, using a special spray gun. Larvae of *Dendrolimus pini*, after hibernating in the soil, cannot climb up to their food in spring and are at the same time exposed to their enemies. Spot treatment usually requires more manpower than indiscriminate, large-scale distribution of pesticides. It will, therefore, be of particular interest in areas where abundant labour is available at low cost.

Local conditions are important in any ecological problem. Therefore, enough energy and money should be put into research adapted to local problems and into education of adequate staff to continue on a large scale what has already been begun in integrated control. Integrated control means more than the invention or imitation of technical tricks to reduce the amount of pesticides used and to consider the conservation and augmentation of biotic regulating agents. For the future, integrated control should be understood as the incorporation of ecologically sound principles of plant protection into the planning and management of crops.

REFERENCES

Campbell, Robert N., and Arthur S. Greathead. 1990. "Control of clubroot of crucifers by liming". p. 90–101. In: A.W. Engelhard (ed.). *Management of Diseases with Macro- and Microelements.* APS Press. American Phytopathological Society, St. Paul, MN.

Ingham, Elaine. 1998. "Replacing methyl bromide with compost". *BioCycle.* December. p. 80–82.

Granatstein, David. 1998. "Suppressing plant diseases with compost". *Good Fruit Grower.* May 1. p. 9–11.

Logsdon, Gene. 1995. *Using compost for plant disease control. Farm Scale Composting.* JG Press, Inc. Emmaus, PA.

Schneider, R.W. (ed.). 1982. *Suppressive Soils and Plant Disease.* The American Phytopathological Society. St. Paul, MN. 88 p.

Trankner, Andreas. 1992. "Use of agricultural and municipal organic wastes to develop suppressiveness to plant pathogens}. p. 35-42. In: E.C. Tjamos, G.C. Papavizas, and R.J. Cook (ed.). *Biological Control of Plant Diseases: Progress and Challenges for the Future.* NATO ASI Series No. 230. Plenum Press, New York, NY.

8

Soil Management for Disease Prevention

Soils are made up of four basic components: minerals, air, water, and organic matter. In most soils, minerals represent around 45% of the total volume, water and air about 25% each, and organic matter from 2% to 5%. The mineral portion consists of three distinct particle sizes classified as sand, silt, or clay. Sand is the largest particle that can be considered soil. Sand is largely the mineral quartz, though other minerals are also present. Quartz contains no plant nutrients, and sand cannot hold nutrients—they leach out easily with rainfall. Silt particles are much smaller than sand, but like sand, silt is mostly quartz.

The smallest of all the soil particles is clay. Clays are quite different from sand or silt, and most types of clay contain appreciable amounts of plant nutrients. Clay has a large surface area resulting from the plate-like shape of the individual particles. Sandy soils are less productive than silts, while soils containing clay are the most productive and use fertilisers most effectively. Soil texture refers to the relative proportions of sand, silt, and clay. A loam soil contains these three types of soil particles in roughly equal proportions. A sandy loam is a mixture containing a larger amount of sand and a smaller amount of clay, while a clay loam contains a larger amount of clay and a smaller amount of sand.

Another soil characteristic—soil structure—is distinct from soil texture. Structure refers to the clumping together or "aggregation" of sand, silt, and clay particles into larger secondary clusters. If you grab a handful of soil, good structure is apparent when the soil crumbles easily in your hand. This is an indication that the sand, silt, and clay particles are aggregated into granules or crumbs. Both texture and structure determine pore space for air and water circulation, erosion resistance, looseness, ease of tillage, and root penetration. While texture is related to the minerals in the soil and does not change with agricultural activities, structure can be improved or destroyed readily by choice and timing of farm practices.

IMPORTANCE OF SOIL ORGANISMS

An acre of living topsoil contains approximately 900 pounds of earthworms, 2,400 pounds of fungi, 1,500 pounds of bacteria, 133 pounds of protozoa, 890 pounds of arthropods and algae, and even small mammals in some cases. Therefore, the soil can be viewed as a living community rather than an inert body.

Soil organic matter also contains dead organisms, plant matter, and other organic materials in various phases of decomposition. Humus, the dark-colored organic material in the final stages of decomposition, is relatively stable. Both organic matter and humus serve as reservoirs of plant nutrients; they also help to build soil structure and provide other benefits. The type of healthy living soil required to support humans now and far into the future will be balanced in nutrients and high in humus, with a broad diversity of soil organisms. It will produce healthy plants with minimal weed, disease, and insect pressure. To accomplish this, we need to work with the natural processes and optimise their functions to sustain our farms.

Considering the natural landscape, you might wonder how native prairies and forests function in the absence of tillage and fertilisers. These soils are tilled by soil organisms, not by machinery. They are fertilised too, but the fertility is used again and again and never leaves the site. Native soils are covered with a layer of plant litter and/ or growing plants throughout the year. Beneath the surface litter, a rich complexity of soil organisms decompose plant residue and dead roots, then release their stored nutrients slowly over time. In fact, topsoil is the most biologically diverse part of the earth. Soil-dwelling organisms release bound-up minerals, converting them into plant-available forms that are then taken up by the plants growing on the site. The organisms recycle nutrients again and again with the death and decay of each new generation of plants. There are many different types of creatures that live on or in the topsoil. Each has a role to play. These organisms will work for the farmer's benefit if we simply manage for their survival. Consequently, we may refer to them as soil livestock. While a great variety of organisms contribute to soil fertility, earthworms, arthropods, and the various microorganisms merit particular attention.

Earthworms

Earthworm burrows enhance water infiltration and soil aeration. Fields that are "tilled" by earthworm tunneling can absorb water at a rate 4 to 10 times that of fields lacking worm tunnels. This reduces water runoff, recharges groundwater, and helps store more soil water for dry spells. Vertical earthworm burrows pipe air deeper into the soil, stimulating microbial nutrient cycling at those deeper levels. When earthworms are present in high numbers, the tillage provided by their burrows can replace some expensive tillage work done by machinery.

Worms eat dead plant material left on top of the soil and redistribute the organic matter and nutrients throughout the topsoil layer. Nutrient-rich organic compounds line their tunnels, which may remain in place for years if not disturbed. During droughts these tunnels allow for deep plant root penetration into subsoil regions of higher moisture content. In addition to organic matter, worms also consume soil and soil microbes. The soil clusters they expel from their digestive tracts are known as *worm casts* or *castings*. These range from the size of a mustard seed to that of a sorghum seed, depending on the size of the worm.

Table 1. Selected nutrient analyses of worm casts compared to those of the surrounding soil.

Nutrient	*Worm casts (Lbs/ac)*	*Soil (Lbs/ac)*
Carbon	171,000	78,500
Nitrogen	10,720	7,000
Phosphorus	280	40
Potassium	900	140

The soluble nutrient content of worm casts is considerably higher than that of the original soil. A good population of earthworms can process 20,000 pounds of topsoil per year—with turnover rates as high as 200 tons per acre having been reported in some exceptional cases. Earthworms also secrete a plant growth stimulant. Reported increases in plant growth following earthworm activity may be partially attributed to this substance, not just to improved soil quality. Earthworms thrive where there is no tillage. Generally, the less tillage the better, and the shallower the tillage the better. Worm numbers can be reduced by as much as 90% by deep and frequent tillage. Tillage reduces earthworm populations by drying the soil, burying the plant residue they feed on, and making the soil more likely to freeze. Tillage also destroys vertical worm burrows and can kill and cut up the worms themselves. Worms are dormant in the hot part of the summer and in the cold of winter. Table 2 shows the effect of tillage and cropping practices on earthworm numbers.

Table 2. Effect of crop management on earthworm populations.

Crop	*Management*	*Worms/foot²*
Corn	Plow	1
Corn	No-till	2
Soybean	Plow	6
Soybean	No-till	14
Bluegrass/clover	—	39
Dairy pasture	—	33

As a rule, earthworm numbers can be increased by reducing or eliminating tillage (especially fall tillage), not using a moldboard plow, reducing residue particle size (using a straw chopper on the combine), adding animal manure, and growing green manure crops. It is beneficial to leave as much surface residue as possible year-round. Cropping systems that typically have the most earthworms are perennial cool-season grass grazed rotationally, warm-season perennial grass grazed rotationally, and annual croplands using no-till. Ridge-till and strip tillage will generally have more earthworms than clean tillage involving plowing and disking. Cool season grass rotationally grazed is highest because it provides an undisturbed environment plus abundant organic matter from the grass roots and fallen grass litter. Generally speaking, worms want their food on top, and they want to be left alone.

Earthworms prefer a near-neutral soil pH, moist soil conditions, and plenty of plant residue on the soil surface. They are sensitive to certain pesticides and some incorporated fertilisers. Carbamate insecticides, including Furadan, Sevin, and Temik, are harmful to earthworms, notes worm biologist Clive Edwards of Ohio State University. Some insecticides in the organophosphate family are mildly toxic to earthworms, while synthetic pyrethroids are harmless to them. Most herbicides have little effect on worms except for the triazines, such as Atrazine, which are moderately toxic. Also, anhydrous ammonia kills earthworms in the injection zone because it dries the soil and temporarily increases the pH there. High rates of ammonium-based fertilisers are also harmful.

Arthropods

In addition to earthworms, there are many other species of soil organisms that can be seen by the naked eye. Among them are sowbugs, millipedes, centipedes, slugs, snails, and springtails. These are the primary decomposers. Their role is to eat and shred the large particles of plant and animal residues. Some bury residue, bringing it into contact with other soil organisms that further decompose it. Some members of this group prey on smaller soil organisms. The springtails are small insects that eat mostly fungi. Their waste is rich in plant nutrients released after other fungi and bacteria decompose it. Also of interest are dung beetles, which play a valuable role in recycling manure and reducing livestock intestinal parasites and flies.

Bacteria

Bacteria are the most numerous type of soil organism: every gram of soil contains at least a million of these tiny one-celled organisms. There are many different species of bacteria, each with its own role in the soil environment. One of the major benefits bacteria provide for plants is in making nutrients available to them. Some species release nitrogen, sulfur, phosphorus, and trace elements from organic matter. Others break down soil minerals, releasing potassium, phosphorus, magnesium, calcium, and iron. Still other species make and release plant growth hormones, which stimulate root growth. Several species of bacteria transform nitrogen from a gas in the air to forms available for plant use, and from these forms back to a gas again. A few species of bacteria fix nitrogen in the roots of legumes, while others fix nitrogen independently of plant association. Bacteria are responsible for converting nitrogen from ammonium to nitrate and back again, depending on certain soil conditions. Other benefits to plants provided by various species of bacteria include increasing the solubility of nutrients, improving soil structure, fighting root diseases, and detoxifying soil.

Fungi

Fungi come in many different species, sizes, and shapes in soil. Some species appear as thread-like colonies, while others are one-celled yeasts. Slime molds and mushrooms are also fungi. Many fungi aid plants by breaking down organic matter or by releasing nutrients from soil minerals. Fungi are generally quick to colonise larger pieces of organic matter and begin the decomposition process. Some fungi produce plant hormones, while others produce antibiotics including penicillin. There are even species of fungi that trap harmful plant-parasitic nematodes. The mycorrhizae are fungi that live either on or in plant roots and act to extend the reach of root hairs into the soil. Mycorrhizae increase the uptake of water and nutrients, especially phosphorus. They are particularly important in degraded or less fertile soils. Roots colonised by mycorrhizae are less likely to be penetrated by root-feeding nematodes, since the pest cannot pierce the thick fungal network. Mycorrhizae also produce hormones and antibiotics that enhance root growth and provide disease suppression. The fungi benefit by taking nutrients and carbohydrates from the plant roots they live in.

Actinomycetes

Actinomycetes are thread-like bacteria that look like fungi. While not as numerous as bacteria, they too perform vital roles in the soil. Like the bacteria, they help decompose organic matter into humus, releasing nutrients. They also produce antibiotics to fight diseases of roots. Many of these same antibiotics are used to treat human diseases. Actinomycetes are responsible for the sweet, earthy smell noticed whenever a biologically active soil is tilled.

Algae

Many different species of algae live in the upper half-inch of the soil. Unlike most other soil organisms, algae produce their own food through photosynthesis. They appear as a greenish film on the soil surface following a saturating rain. Algae improve soil structure by producing slimy substances that glue soil together into water-stable aggregates. Some species of algae (the blue-greens) can fix their own nitrogen, some of which is later released to plant roots.

Protozoa

Protozoa are free-living microorganisms that crawl or swim in the water between soil particles. Many soil protozoa are predatory, eating other microbes. One of the most common is an amoeba that eats bacteria. By eating and digesting bacteria, protozoa speed up the cycling of nitrogen from the bacteria, making it more available to plants.

Nematodes

Nematodes are abundant in most soils, and only a few species are harmful to plants. The harmless species eat decaying plant litter, bacteria, fungi, algae, protozoa, and other nematodes. Like other soil predators, nematodes speed the rate of nutrient cycling.

SOIL QUALITY

All these organisms—from the tiny bacteria up to the large earthworms and insects—interact with one another in a multitude of ways in the soil ecosystem. Organisms not directly involved in decomposing plant wastes may feed on each other or each other's waste products or the other substances they release. Among the substances released by the various microbes are vitamins, amino acids, sugars, antibiotics, gums, and waxes.

Roots can also release into the soil various substances that stimulate soil microbes. These substances serve as food for select organisms. Some scientists and practitioners theorize that plants use this means to stimulate the specific population of microorganisms capable of releasing or otherwise producing the kind of nutrition needed by the plants.

Research on life in the soil has determined that there are ideal ratios for certain key organisms in highly productive soils. The Soil Foodweb Lab, located in Oregon, tests soils and makes fertility recommendations that are based on this understanding. Their goal is to alter the makeup of the soil microbial community so it resembles that of a highly fertile and productive soil.

Because we cannot see most of the creatures living in the soil and may not take time to observe the ones we can see, it is easy to forget about them. Table 3 for estimates of typical amounts of various organisms found in fertile soil.

Table 3. Weights of soil organisms in the top 7 inches of fertile soil.

Organism	*Pounds of liveweight/acre*
Bacteria	1000
Actinomycetes	1000
Molds	2000
Algae	100
Protozoa	200
Nematodes	50
Insects	100
Worms	1000
Plant roots	2000

ORGANIC MATTER

Understanding the role that soil organisms play is critical to sustainable soil management. Based on that understanding, focus can be directed toward strategies that build both the numbers and the diversity of soil organisms. Like cattle and other farm animals, soil livestock require proper feed. That feed comes in the form of organic matter.

Organic matter and humus are terms that describe somewhat different but related things. Organic matter refers to the fraction of the soil that is composed of both living organisms and once-living residues in various stages of decomposition. Humus is only a small portion of the organic matter. It is the end product of organic matter decomposition and is relatively stable. Further decomposition of humus occurs very slowly in both agricultural and natural settings. In natural systems, a balance is reached between the amount of humus formation and the amount of humus decay. This balance also occurs in most agricultural soils, but often at a much lower level of soil humus. Humus contributes to well-structured soil that, in turn, produces high-quality plants. It is clear that management of organic matter and humus is essential to sustaining the whole soil ecosystem.

The benefits of a topsoil rich in organic matter and humus are many. They include rapid decomposition of crop residues, granulation of soil into water-stable aggregates, decreased crusting and clodding, improved internal drainage, better water infiltration, and increased water and nutrient holding capacity. Improvements in the soil's physical structure facilitate easier tillage, increased water storage capacity, reduced erosion, better formation and harvesting of root crops, and deeper, more prolific plant root systems.

Soil organic matter can be compared to a bank account for plant nutrients. Soil containing 4% organic matter in the top seven inches has 80,000 pounds of organic matter per acre. That 80,000 pounds of organic matter will contain about 5.25% nitrogen, amounting to 4,200 pounds of nitrogen per acre. Assuming a 5% release rate during the growing season, the organic matter could supply 210 pounds of nitrogen to a crop. However, if the organic matter is allowed to degrade and lose nitrogen, purchased fertiliser will be necessary to prop up crop yields.

All the soil organisms mentioned previously, except algae, depend on organic matter as their food source. Therefore, to maintain their populations, organic matter must be renewed from plants growing on the soil, or from animal manure, compost, or other materials imported from off site. When soil livestock are fed, fertility is built up in the soil, and the soil will feed the plants.

Ultimately, building organic matter and humus levels in the soil is a matter of managing the soil's living organisms—something akin to wildlife management or animal husbandry. This entails working to maintain favorable conditions of moisture, temperature, nutrients, pH, and aeration. It also involves providing a steady food source of raw organic material.

Soil Tilth

A soil that drains well, does not crust, takes in water rapidly, and does not make clods is said to have good tilth. Tilth is the physical condition of the soil as it relates to tillage ease, seedbed quality, easy seedling emergence, and deep root penetration. Good tilth is dependent on aggregation—the process whereby individual soil particles are joined into clusters or "aggregates."

Aggregates form in soils when individual soil particles are oriented and brought together through the physical forces of wetting and drying or freezing and thawing. Weak electrical forces from calcium and magnesium hold soil particles together when the soil dries. When these aggregates become wet again, however, their stability is challenged, and they may break apart. Aggregates can also be held together by plant roots, earthworm activity, and by glue-like products produced by soil microorganisms. Earthworm-created aggregates are stable once they come out of the worm. An aggregate formed by physical forces can be bound together by fine root hairs or threads produced by fungi.

Crusting is a common problem on soils that are poorly aggregated. Crusting results chiefly from the impact of falling raindrops. Rainfall causes clay particles on the soil surface to disperse and clog the pores immediately beneath the surface. Following drying, a sealed soil surface results in which most of the pore space has been drastically reduced due to clogging from dispersed clay particles. Subsequent rainfall is much more likely to run off than to flow into the soil.

Since raindrops start crusting, any management practices that protect the soil from their impact will decrease crusting and increase water flow into the soil. Mulches and cover crops serve this purpose well, as do no-till practices, which allow the accumulation of surface residue. Also, a well-aggregated soil will resist crusting because the water-stable aggregates are less likely to break apart when a raindrop hits them.

Long-term grass production produces the best-aggregated soils. A grass sod extends a mass of fine roots throughout the topsoil, contributing to the physical processes that help form aggregates. Roots continually remove water from soil microsites, providing local wetting and drying effects that promote aggregation. Fine root hairs also bind soil aggregates together.

Roots also produce food for soil microorganisms and earthworms, which in turn generate compounds that bind soil particles into water-stable aggregates. In addition, perennial grass sods provide protection from raindrops and erosion. Thus, a perennial cover creates a combination of conditions optimal for the creation and maintenance of well-aggregated soil.

Conversely, cropping sequences that involve annual plants and extensive cultivation provide less vegetative cover and organic matter, and usually result in a rapid decline in soil aggregation.

Farming practices can be geared to conserve and promote soil aggregation. Because the binding substances are themselves susceptible to microbial degradation, organic matter needs to be replenished to maintain microbial populations and overall aggregated soil status. Practices should conserve aggregates once they are formed, by minimising factors that degrade and destroy aggregation. Some factors that destroy or degrade soil aggregates are:

- bare soil surface exposed to the impact of raindrops
- removal of organic matter through crop production and harvest without return of organic matter to the soil
- excessive tillage
- working the soil when it is too wet or too dry
- use of anhydrous ammonia, which speeds up decomposition of organic matter
- excess nitrogen fertilisation
- allowing the build-up of excess sodium from irrigation or sodium-containing fertilisers

Tillage

Several factors affect the level of organic matter that can be maintained in a soil. Among these are organic matter additions, moisture, temperature, tillage, nitrogen levels, cropping, and fertilisation. The level of organic matter present in the soil is a direct function of how much organic material is being produced or added to the soil versus the rate of decomposition. Achieving this balance entails slowing the speed of organic matter decomposition, while increasing the supply of organic materials produced on site and/or added from off site.

Moisture and temperature also profoundly affect soil organic matter levels. High rainfall and temperature promote rapid plant growth, but these conditions are also favorable to rapid organic matter decomposition and loss. Low rainfall or low temperatures slow both plant growth and organic matter decomposition. The native Midwest prairie soils originally had a high amount of organic matter from the continuous growth and decomposition of perennial grasses, combined with a moderate temperature that did not allow for rapid decomposition of organic matter. Moist and hot tropical areas may appear lush because of rapid plant growth, but soils in these areas are low in nutrients. Rapid decomposition of organic matter returns nutrients back to the soil, where they are almost immediately taken up by rapidly growing plants.

Tillage can be beneficial or harmful to a biologically active soil, depending on what type of tillage is used and when it is done. Tillage affects both erosion rates and soil organic matter decomposition rates. Tillage can reduce the organic matter level in croplands below 1%, rendering them biologically dead. Clean tillage involving moldboard plowing and disking breaks down soil aggregates and leaves the soil prone to erosion from wind and water. The moldboard plow can bury crop residue and topsoil to a depth of 14 inches.

At this depth, the oxygen level in the soil is so low that decomposition cannot proceed adequately. Surface-dwelling decomposer organisms suddenly find themselves suffocated and soon die. Crop residues that were originally on the surface but now have been turned under will putrefy in the oxygen-deprived zone. This rotting activity may give a putrid smell to the soil. Furthermore, the top few inches of the field are now often covered with subsoil having very little organic matter content and, therefore, limited ability to support productive crop growth.

The topsoil is where the biological activity happens—it's where the oxygen is. That's why a fence post rots off at the surface. In terms of organic matter, tillage is similar to opening the air vents on a wood-burning stove; adding organic matter is like adding wood to the stove. Ideally, organic matter decomposition should proceed as an efficient burn of the "wood" to release nutrients and carbohydrates to the soil organisms and create stable humus. Shallow tillage incorporates residue and speeds the decomposition of organic matter by adding oxygen that microbes need to become more active.

In cold climates with a long dormant season, light tillage of a heavy residue may be beneficial; in warmer climates it is hard enough to maintain organic matter levels without any tillage.

Tillage also reduces the rate of water entry into the soil by removal of ground cover and destruction of aggregates, resulting in compaction and crusting. Table 4 shows three different tillage methods and how they affect water entry into the soil. Notice the direct relationship between tillage type, ground cover, and water infiltration. No-till has more than three times the water infiltration of the moldboard-plowed soil. Additionally, no-till fields will have higher aggregation from the organic matter decomposition on site. The surface mulch typical of no-till fields acts as a protective skin for the soil. This soil skin reduces the impact of raindrops and buffers the soil from temperature extremes as well as reducing water evaporation.

Table 4. Tillage effects on water infiltration and ground cover.

	Water Infiltration mm/minute	*Ground Cover Percent*
No-till	2.7	48
Chisel Plow	1.3	27
Moldboard Plow	0.8	12

Both no-till and reduced-tillage systems provide benefits to the soil. The advantages of a no-till system include superior soil conservation, moisture conservation, reduced water runoff, long-term buildup of organic matter, and increased water infiltration. A soil managed without tillage relies on soil organisms to take over the job of plant residue incorporation formerly done by tillage. On the down side, no-till can foster a reliance on herbicides to control weeds and can lead to soil compaction from the traffic of heavy equipment.

Pioneering development work on chemical-free no-till farming is proceeding at several research stations and farms in the eastern U.S. Pennsylvania farmer Steve Groff has been farming no-till with minimal or no herbicides for several years. Groff grows cover crops extensively in his fields, rolling them down in the spring using a 10-foot rolling stalk chopper. This rolling chopper kills the rye or vetch cover crop and creates a nice no-till mulch into which he plants a variety of vegetable and grain crops. After several years of no-till production, his soils are mellow and easy to plant into.

Other conservation tillage systems include ridge tillage, minimum tillage, zone tillage, and reduced tillage, each possessing some of the advantages of both conventional till and no-till. These systems represent intermediate tillage systems, allowing more flexibility than either a no-till or conventional till system might. They are more beneficial to soil organisms than a conventional clean-tillage system of moldboard plowing and disking.

Adding manure and compost is a recognised means for improving soil organic matter and humus levels. In their absence, perennial grass is the only crop that can regenerate and increase soil humus. Cool-season grasses build soil organic matter faster than warm-season grasses because they are growing much longer during a given year. When the soil is warm enough for soil organisms to decompose organic matter, cool-season grass is growing. While growing, it is producing organic matter and cycling minerals from the decomposing organic matter in the soil. In other words, there is a net gain of organic matter because the cool-season grass is producing organic matter faster than it is being used up. With warm-season grasses, organic matter production during the growing season can be slowed during the long dormant season from fall through early spring. During the beginning and end of this dormant period, the soil is still biologically active, yet no grass growth is proceeding. Some net accumulation of organic matter can occur under warm-season grasses, however. In a Texas study, switchgrass grown for four years increased soil carbon content from 1.1% to 1.5% in the top 12 inches of soil. In hot and moist regions, a cropping rotation that includes several years of pasture will be most beneficial.

Effect of Nitrogen

Excessive nitrogen applications stimulate increased microbial activity, which in turn speeds organic matter decomposition. The extra nitrogen narrows the ratio of carbon to nitrogen in the soil. Native or uncultivated soils have approximately 12 parts of carbon to each part of nitrogen, or a C:N ratio of 12:1. At this ratio, populations of decay bacteria are kept at a stable level, since additional growth in their population is limited by a lack of nitrogen. When large amounts of inorganic nitrogen are added, the C:N ratio is reduced, which allows the populations of decay organisms to explode as they decompose more organic matter with the now abundant nitrogen.

While soil bacteria can efficiently use moderate applications of inorganic nitrogen accompanied by organic amendments (carbon), excess nitrogen results in decomposition of existing organic matter at a rapid rate. Eventually, soil carbon content may be reduced to a level where the bacterial populations are on a starvation diet. With little carbon available, bacterial populations shrink, and less of the free soil nitrogen is absorbed. Thereafter, applied nitrogen, rather than being cycled through microbial organisms and re-released to plants slowly over time, becomes subject to leaching. This can greatly reduce the efficiency of fertilisation and lead to environmental problems.

To minimise the fast decomposition of soil organic matter, carbon should be added with nitrogen. Typical carbon sources—such as green manures, animal manure, and compost—serve this purpose well. Amendments containing too high a carbon to nitrogen ratio (25:1 or more) can tip the balance the other way, resulting in nitrogen being tied up in an unavailable form. Soil organisms consume all the nitrogen in an effort to decompose the abundant carbon; tied up in the soil organisms, nitrogen remains unavailable for plant uptake. As soon as a soil microorganism dies and decomposes, its nitrogen is consumed by another soil organism, until the balance between carbon and nitrogen is achieved again.

Conventional Fertilisers

Commercial fertiliser can be a valuable resource to farmers in transition to a more sustainable system and can help meet nutrient needs during times of high crop nutrient demand or when weather conditions result in slow nutrient release from organic resources. Commercial fertilisers have the advantage of supplying plants with immediately available forms of nutrients. They are often less expensive and less bulky to apply than many natural fertilisers.

Not all conventional fertilisers are alike. Many appear harmless to soil livestock, but some are not. Anhydrous ammonia contains approximately 82% nitrogen and is applied subsurface as a gas. Anhydrous speeds the decomposition of organic matter in the soil, leaving the soil more compact as a result. The addition of anhydrous causes increased acidity in the soil, requiring 148 pounds of lime to neutralise 100 pounds of anhydrous ammonia, or 1.8 pounds of lime for every pound of nitrogen contained in the anhydrous. Anhydrous ammonia initially kills many soil microorganisms in the application zone. Bacteria and actinomycetes recover within one to two weeks to levels higher than those prior to treatment. Soil fungi, however, may take seven weeks to recover. During the recovery time, bacteria are stimulated to grow more, and decompose more organic matter, by the high soil nitrogen content. As a result, their numbers increase after anhydrous applications, then decline as available soil organic matter is depleted. Farmers commonly report that the long-term use of synthetic fertilisers, especially anhydrous ammonia, leads to soil compaction and poor tilth. When bacterial populations and soil organic matter decrease, aggregation declines, because existing glues that stick soil particles together are degraded, and no other glues are being produced.

Potassium chloride (KCl) (0-0-60 and 0-0-50), also known as muriate of potash, contains approximately 50 to 60% potassium and 47.5% chloride. Muriate of potash is made by refining potassium chloride ore, which is a mixture of potassium and sodium salts and clay from the brines of dying lakes and seas. The potential harmful effects from KCl can be surmised from the salt concentration of the material. Table 5 shows that, pound for pound, KCl is surpassed only by table salt on the salt index. Additionally, some plants such as tobacco, potatoes, peaches, and some legumes are especially sensitive to chloride. High rates of KCl must be avoided on such crops. Potassium sulfate, potassium nitrate, sul-po-mag, or organic sources of potassium may be considered as alternatives to KCl for fertilisation.

Sodium nitrate, also known as Chilean nitrate or nitrate of soda, is another high-salt fertiliser. Because of the relatively low nitrogen content of sodium nitrate, a high amount of sodium is added to the soil when normal applications of nitrogen are made with this material. The concern is that excessive sodium acts as a dispersant of soil particles, degrading aggregation. The salt index for KCl and sodium nitrate can be seen in Table 5.

Table 5. Salt index for various fertilisers.

Material	*Salt Index*	*Salt index per unit of plant food*
Sodium chloride	153	2.90
Potassium chloride	116	1.90
Ammonium nitrate	105	3.00
Sodium nitrate	100	6.10
Urea	75	1.60
Potassium nitrate	74	1.60
Ammonium sulfate	69	3.30
Calcium nitrate	53	4.40
Anhydrous ammonia	47	0.06
Sulfate-potash-magnesia	43	2.00
Di-ammonium phosphate	34	1.60
Monammonium phosphate	30	2.50
Gypsum	8	0.03
Calcium carbonate	5	0.01

Topsoil

Topsoil is the capital reserve of every farm. Ever since mankind started agriculture, erosion of topsoil has been the single largest threat to a soil's productivity—and, consequently, to farm profitability. This is still true today. In the U.S., the average acre of cropland is eroding at a rate of 7 tons per year. To sustain agriculture means to sustain oil resources, because that's the source of a farmer's livelihood. The major productivity osts to the farm associated with soil erosion come from the replacement of lost nutrients nd reduced water holding ability, accounting for 50 to 75% of productivity loss. Soil hat is removed by erosion typically contains about three times more nutrients than the oil left behind and is 1.5 to 5 times richer in organic matter. This organic matter loss ot only results in reduced water holding capacity and degraded soil aggregation, but lso loss of plant nutrients, which must then be replaced with nutrient amendments.Five ons of topsoil (the so-called tolerance level) can easily contain 100 pounds of nitrogen,) pounds of phosphate, 45 pounds of potash, 2 pounds of calcium, 10 pounds of agnesium, and 8 pounds of sulfur.

Water erosion gets started when falling rainwater collides with bare ground and etaches soil particles from the parent soil body. After enough water builds up on the oil surface, following detachment, overland water flow transports suspended soil down-ope. Suspended soil in the runoff water abrades and detaches additional soil particles s the water travels overland. Preventing detachment is the most effective point of rosion control because it keeps the soil in place. Other erosion control practices seek) slow soil particle transport and cause soil to be deposited before it reaches streams. hese methods are less effective at protecting the quality of soil within the field.

Commonly implemented practices to slow soil transport include terraces and iversions. Terraces, diversions, and many other erosion "control" practices are largely nnecessary if the ground stays covered year-round. For erosion prevention, a high ercentage of ground cover is a good indicator of success, while bare ground is an "early arning" indicator for a high risk of erosion. Muddy runoff water and gullies are "too-te" indicators. The soil has already eroded by the time it shows up as muddy water, d it's too late to save soil already suspended in the water. Protecting the soil from osion is *the* first step toward a sustainable agriculture. Since water erosion is initiated raindrop impact on bare soil, any management practice that protects the soil from indrop impact will decrease erosion and increase water entry into the soil. Mulches, ver crops, and crop residues serve this purpose well.

Additionally, well-aggregated soils resist crusting because water-stable aggregates are ss likely to break apart when the raindrop hits them. Adequate organic matter with gh soil biological activity leads to high soil aggregation.

Many studies have shown that cropping systems that maintain a soil-protecting plant nopy or residue cover have the least soil erosion. This is universally true. Long-term opping studies begun in 1888 at the University of Missouri provide dramatic evidence this. Gantser and colleagues examined the effects of a century of cropping on soil osion. They compared depth of topsoil remaining after 100 years of cropping. As the table shows, the cropping system that maintained the highest amount of permanent ground cover had the greatest amount of topsoil left.

Table 6. Topsoil depth remaining after 100 years of different cropping practices.

Crop Sequence	*Inches of topsoil remaining*
Continuous Corn	*7.7*
6-year rotation	12.2
Continuous timothy grass	17.4

The researchers commented that subsoil had been mixed with topsoil in the continuous corn plots from plowing, making the real topsoil depth less than was apparent. In reality, all the topsoil was lost from the continuous corn plots in only 100 years. The rotation lost about half the topsoil over 100 years. How can we feed future generations with this type of farming practice? Some soils naturally have very thick topsoil, while other soils have thin topsoil over rock or gravel. Roughly 8 tons/acre/year of soil-erosion loss amounts to the thickness of a dime spread over an acre. Twenty dimes stack up to 1-inch high. So a landscape with an 8-ton erosion rate would lose an inch of topsoil about every 20 years. On a soil with a thick topsoil, this amount is barely detectable within a person's lifetime and may not be noticed. Soils with naturally thin topsoils or topsoils that have been previously eroded can be transformed from productive to degraded land within a generation.

Tillage for the production of annual crops is the major problem in agriculture, causing soil erosion and the loss of soil quality. Any agricultural practice that creates and maintains bare ground is inherently less sustainable than practices that keep the ground covered throughout the year. Wes Jackson has spent much of his career developing perennial grain crops and cropping systems that mimic the natural prairie. Perennial grain crops do not require tillage to establish year after year, and the ground is left covered. Ultimately, this is the future of grain production and truly represents a new vision for how we produce food.

INTEGRATED SOIL MANAGEMENT

Food security is the most important factor that determines the survival of human kind. Without food security, a nation cannot expect better life for its people. Famines in India are "a nightmare of the past". The green revolution witnessed in late 1960s has contributed immensely over the years to cereal production in India and hence a substantial increase in the net per capita availability of food grains was registered (Table 7). This has led to a nationwide sense of complacency that, in a way, slowed down the growth rate in agricultural production during 1990s, while the population continued to grow at a high rate. The net result was a decline in the per capita food grain availability in the terminal decade of 20th Century. Even with present level of production, there is enough food in the country to meet energy and protein requirements of the current population, if the food were distributed equitably according to needs. But as we see, surplus production and widespread hunger coexist at the national level. At present, India alone accounts for one fourth of all world hunger. It is particularly ironic that there are 200 million food-insecure people in a country that currently bas buffer stocks of food grains in excess of 60 million metric tonnes.

Table 7. Per capita net availability of food grains in India (g/day)

Year	*Cereals*	*Pulses*	*Total food grains*
1951	334.2	60.7	394.9
1961	399.7	69.0	468.7
1971	417.6	51.2	468.8
1981	417.3	37.5	454.8
1991	468.5	41.6	510.1
2001	385.1	29.1	414.1

Inadequate or lack of purchasing power among the poor is the main cause of food insecurity in rural India. As reported by Rajendra Prasad, the per capita consumption of most food items in rural India is far below the recommended dietary allowances. Though the per capita intake of cereals in all regions, and sugar and milk consumption in North and Western regions is closer to or above the standard requirements, the consumption of all other food items throughout the country is woefully lower than their respective dietary requirements as per ICMR (Indian Council of Medical Research) norms.

A general low intake of pulses, vegetables, fruits, fats and oils, eggs, meat and fish is responsible for widespread occurrence of protein energy malnutrition (PEM) and chronic energy deficiency (CED). It was reported that 23 to 70 percent of the rural population in different parts of the country is suffering from protein energy malnutrition, while the chronic energy deficiency affected 17 to 54 percent of people (Table 8). Prevalence of poverty and low and fluctuating income levels also limit the access to diversified diet and thus adversely affect balanced diet. The vegetable products account for a lion share in the intake of all dietary constituents. A comparison of share of vegetable products and animal products in meeting total dietary energy, protein and fat in India, USA and the World as a whole makes this point clear. In India, vegetable products provide 93 percent dietary energy, 84 percent protein and 73 percent fat, whil animal products supply the remaining small portion, i.e., 7 percent, 16 percent and 2 percent of energy, protein and fat, respectively. On the contrary, in a developed countr like USA, the animal products account for 30 percent, 64 percent and 511 percent shar in meeting dietary energy, protein and fat supply, respectively. Child malnutrition rate in India are still very high. According to the UNDP, 53 percent of children under fiv in India were under-weight during the period 1990-97, the highest rate from any c the 174 developing countries listed.

Table 8. Extent of PEM and CED in rural India

Region	*Percent of population with*	
	Protein energy malnutrition (PEM)	*Chronic energy deficiency (CED)*
Northern	34.9-36.9	23.0-44.0
Eastern and Central	23.5-58.2	17.1-57.3
Western	30.2-39.8	36.2-53.1
South	31.4-70.3	33.2-53.8

In India, unabated growth in population has been and will continue to be the sing most factors that have the potential to negate all the progress made in agricultur production. India's population grew at an annual growth rate of around 2 percent 1970s, 1980s and 1990s to reach 1 027 million in 2001 and is estimated to increase furth to 1 262 and 1 542 million by the year 2011 and 2021, respectively. Growing populatic means mounting more pressure on natural resources to meet increased food deman According to a conservative estimate, the food grain demand in India for the years 20 and 2020 is projected to be 246 and 294 mt, respectively (Table 9). This means th India's food grain production has to increase from 212 mt to 246 mt in 2010 and the to 294 mt in 2020. It is by all means a daunting task and the ability to accomplish th task determines the future food security of the country.

Table 9. Current production and future demands of food grains in India

Food item	*Current (2001-02) production (mt)*	*Estimated demand (mt)*	
		2010	*2020*
Rice	93.1	3.6	122.1
Wheat	71.8	85.8	102.8
Total cereals	198.8	224.4	265.8
Pulses	13.2	21.4	27.8
Total food grains	212.0	245.8	293.6

With continued rise in population, the arable land to man ratio has decreased from 0.5 ha to 0.14 ha at present and is expected to decline further to 0.08 ha by year 2020. The average number of land holdings has also increased simultaneously from 77 million to over 115 million at present due to population growth and the law of inheritance of land property. The average size of operational farm holding is only 1.57 ha. Further, about 78 percent of the 115 million farm holders in the country come under small and marginal category with the size of farm being less than 2 ha. The small size and scattered nature of the holdings will adversely affect the farm efficiency and will result in high cost of production. This in turn will result in low productivity and thus reduced agricultural sustainability and food security.

The total factor productivity (TFP) is used as an important measure to evaluate the performance of a production system and sustainability of its growth pattern. As stated earlier, adoption of green revolution technology led to a phenomenal growth in agricultural production during 1970s and 1980s. But of late, there are signs of fatigue in the agricultural growth process. In spite of continued growth of inputs, there has been no matching growth in agricultural production during 1990s, indicating a decrease in TEP. The declining trends of annual growth rate of productivity in respect of all major crops (Table 10) are also suggestive of decreasing TFP in Indian agriculture.

Table 10. Productivity growth rate of important crops in India

Crop	*Annual growth rate in productivity (%)*		
	1980-81 to 1989-90	*1990-91 to1999-2000*	*2000-01 to2002-03*
Rice	3.19	1.27	-0.72
Wheat	3.10	2.11	0.73
Pulses	1.61	0.96	-1.84
Total food grains	2.74	1.52	-0.69
Oilseeds	2.43	1.25	-3.83
Non-food grains	2.31	1.04	-1.02
All principal crops	2.56	1.31	-0.87

In fact, all the crops except wheat registered a negative annual growth rate in their productivity during the recent past. If this alarming trend is allowed to continue, it will spell doom on the country's future food security prospects. Reasons for decreasing the total factor productivity are:

(1) High nutrient turn over in soil-plant system coupled with low and imbalanced fertiliser use,

(2) Emerging deficiencies of micro and secondary nutrients (S, Zn, B, Fe, Mn, etc.),

(3) Soil degradation due to acidification, aluminum toxicity, soil salinisation and alkalisation, soil erosion,

(4) Wide nutrient gap between nutrient demand and supply, and

(5) Consequent deterioration in soil physical, biological and chemical quality and low fertiliser use efficiency.

The growth in fertiliser consumption slowed down during 1990's and there is stagnation in consumption during the last 4-5 years. After achieving a record consumption level of 18.1 mt of NPK in 1999-2000, the total NPK consumption is hovering around 16-17 mt during the last 3 years. At the present level of crop production, there exists a negative balance of 10 mt between nutrient (NPK) demand by crops and supply of nutrient through application of fertiliser annually.

The stagnant situation in fertiliser consumption and higher negative nutrient balance are posing a threat to soil quality and sustainable agriculture. It is now imperative to review the reasons for the stagnant trend in fertiliser consumption and take remedial action to alter this trend. The stagnant trend in fertiliser consumption despite slow increase in maximum retail price of fertilisers reveals that besides pricing, there are various other reasons which affect the fertiliser consumption. Weather cannot be solely blamed for the stagnant situation as the performance of southwest monsoon had been normal in the last few years. The total NPK consumption did not exceed 17 mt during the year-2003-04, in spite of having good southwest monsoon rainfall. Deteriorating soil quality and the emerging deficiencies in secondary and micronutrients aside from major nutrients appear to be one of the major factors in the stagnation of fertiliser consumption. A cereal production of 5-10 t/ha/year in rice-wheat rotation, which is the backbone of India's food security removes 380-760 kg N-P_2O_5-K_2O per hectare per year. Farmers generally apply 50 percent to 80 percent of this amount. Thus there is a gradual depletion of the inherent soil fertility.

Soil Nutrient Balances

There are hardly any farm-level exercises which have been or are being conducted to monitor nutrient balances in intensively cropped areas. It is generally accepted that soils are being mined and that their nutrient capitals being continuously depleted throughout intensively cultivated areas. No quantitative or semi-quantitative estimates, however, are available on nutrient recycling or balances based on various input-output components at the farm level. This is an area where some insights from a few well-defined benchmark farms (not research stations) will be extremely valuable in developing sustainable systems, not only for site-specific adoption, but also for adoption to similar environments.

The fate of soil nutrient capital and balance in the two most important cropping systems of Uttar Pradesh is illustrated below (Table 11). These computations are primarily illustrative and based on several assumptions, due to the present inadequate database. The initial soil nutrient capital is taken to reflect the soil's low status in nitrogen, medium in phosphorus, and high in potassium for the plow layer. Fertiliser inputs for the sugarcane-wheat system are typical of the practice.

Table 11. Nutrient balance after a sugarcane-wheat system in western Uttar Pradesh (productivity 120 mt cane/ha/2 crops + 3 tonnes wheat grain/ha)

Item	*N*	P_2O_5	K_2O
Initial available soil nutrient capital (kilograms per hectare)	280	40	336
For sugarcane plant crop			
Fertiliser input (kg/ha)125	58	10	
10 t/ha FYM (0.75-0.175-0.55) of N-P_2O_5-K_2O	75	18	55
1 t/ha press mud (0.026-1.70-0.24% available N-P_2O_5-K_2O)	1	17	2
Green manure (not practiced)	0	0	0
Crop residues (not recycled)	0	0	0
Total nutrient capital	481	133	403
Nutrient uptake by 60 t/ha sugarcane crop (kg/ha)	135	30	144
Losses from soil (25% of fertiliser N)	31	0	0
Nutrient balance after cane harvest (kg/ha)	315	103	215
For sugarcane ratoon crop			
Starting soil nutrient capital	315	103	215
Fertiliser input (kg/ha)	62	29	5
3 t/ha cane residues recycled (0.4-0.18-1.28)	12	5	38
FYM not used	0	0	0
Total nutrient capital	389	137	258
Nutrient uptake by 60 t/ha ratoon crop (kg/ha)	135	30	188
Losses from soil (25% of fertiliser N)	16	0	0
Nutrient balance after ratoon	238	107	70
For wheat crop			
Starting soil nutrient capital (kg/ha)	238	107	70
Fertiliser input	100	50	10
FYM not used	0	0	0
Crop residues burnt	0	0	38
Total nutrient capital	338	157	118
Nutrient uptake by 4 t/ha wheat crop (kg/ha)	96	36	132
Losses from soil (25% of fertiliser N)	25	0	0

Nutrient balance after wheat (kg/ha)	217	121	-14
Change in initial soil capital	-63	81	-350
Percentage change	-23	102	-104
Initial soil nutrient capital	280	40	336
Capital after sugarcane plant crop	315	103	215
Capital after sugarcane ratoon			
Capital after wheat	217	121	-14
Capital after the system	217	121	-14
Change after 2 years (one crop cycle) in percent	-63(-23%)	81(102%)	-350 (-104%)

The analysis shows that after one cycle of the sugarcane-wheat system, the initial soil nutrient capital decreased by 23 percent in the case of N, increased by 1.2 percent in the case of P but decreased by 104 percent in the case of K. The improvement in P status was attributed to its application to both the main crops and input of FYM and press mud, that less was removed from the crop than was added and the ability of P (unlike N) to accumulate in the soil. The large depletion in K was due to its very weak position in the fertiliser use pattern and crop removal exceeding the K input.

The apparent K balance of long-term fertiliser experiment under maize-wheat-cowpea during 27 years of cropping showed that mining of soil K occurred even under NPK and NPK + FYM treatments, i.e. application of 15 t FYM/ha along with recommended rates of NPK. This shows that the selection of suitable components of INM should vary with cropping systems and nutrient requirement. Integration of crop residues, along with farmyard manure and fertilisers, may arrest the mining of K from soils where the production systems have higher K demand.

High Nutrient Turnover in Soil-plant System

Fertiliser consumption in India is grossly imbalanced since the beginning. It is tilted more towards N followed by P. Further the decontrol of the phosphatic and potassic fertilisers resulted in more than doubling the prices of phosphatic and potassic fertilisers. Thus, the already unbalanced consumption ratio of 6:2.4:1 ($N:P_2O_5:K_2O$) in 1990-91 has widened to 7:2.7:1 in 2000-01 as against favourable ratio of 4:2:1 implying there from that farmers started adding more nitrogen and proportionately less phosphatic and potassic fertilisers. Even today, the situation is grim as far as fertiliser application by farmers is concerned. In many areas the imbalanced fertilisation is the root cause of poor crop yields and poor soil fertility status.

Accordingly, agro-ecological regions 4, 9 14, 15 and 18 and cropping systems like rice-wheat, maize-wheat, rice-pulse, potato-wheat and sugarcane demands immediate attention to correct the imbalances in nutrient consumptions to prevent further deterioration of soil quality and to break the yield barriers. There is wide variation in the consumption ratios of fertilisers from region to region but in the absence of information on the extent of cultivated area and details of cropping patterns in each agro-ecological region, it is difficult to estimate the crop removal of each region.

Deficiencies of Secondary and Micronutrient in Soils

Intensive cropping systems are heavy feeders and are bound to heavily extract nutrients from the soil. Hence, nutrient deficiencies are inevitable unless steps are taken to restore fertility levels. Deficiencies of essential elements in Indian soils and crops started emerging during the 1950s after the initiation of the government of Independent India, a five-year plan to give a fillip to food production through intensification. As food production increased with time, the number of elements becoming deficient in soils and crops also increased.

Unless corrective measures are taken immediately, the list of essential elements becoming deficient is bound to increase further. A classical example of the effect of imbalanced fertiliser use is the fertility decline in an intensive cropping for over 25 years that has been reported by Swarup and Ganeshamurthy. When only N was applied the P and K status in soils at all the centers have gone down. When N & P were applied the soil K status declined more conspicuously in alluvial soils (Ludhiana), Terai soils (Pantnagar) and laterite soils (Bhubaneshwar). All secondary and micronutrients generally declined in all the soils.

Although not universal, deficiency of S has cropped up as serious obstacle in the sustainability of yields in cropping systems particularly if a sulphur responsive crop like rice, oil seed or pulse crop is involved. The extent of S problems depends more on input of S through irrigation and atmosphere, the information from which is completely lacking. The results of long-term experiments (Table 12) show that response was very little in certain crops like wheat and jute and very conspicuous in certain other crops like rice and soybean.

Micronutrient deficiencies in soils are also emerging as yield limiting factors. Analysis of 1.5 lakh soil samples from different regions of the country indicated that about 47 percent of soils were deficient in available Zn, 20 percent of samples were deficient in available B, 18 percent of samples were deficient in Mo, 12 percent of samples were deficient in available Fe and 5 percent of soil samples were found deficient in available Cu. Among the micronutrients, zinc deficiency is widely encountered followed by B, Mo and Fe, in that order.

Table 12. Mean grain yield response (kg/ha) of rice and wheat at Pantnagar

Nutrient or FYM	*Mean response over 5 years (1987-92)*		*Mean response over 20 years (1972-92)*	
	Rice	*Wheat*	*Rice*	*Wheat*
N	1 864	2 372	1 512	2 140
P	124	213	-16	47
K	211	109	430	71
S	183	184	261	150
Zn	520	543	285	307
FYM	587	645	745	623

Field scale deficiency of Zn in crops is being increasingly reported. But suggestions that B and Mo as yield limiting factors are not convincing as trials that include these elements rarely generate conclusive evidence to support this hypothesis. Field scale Mo deficiencies and Mn as a factor of yield decline is not common. However, exception to this is in the rice-wheat system on sandy soils and reclaimed sodic soil. Continuous cropping of rice-wheat on these soils led to deficiency of Mn in wheat crop following leaching of reduced Mn from surface soils under rice culture. The productivity of wheat could be restored by soil and foliar applications of Mn.

SOIL DEGRADATION

It has been stated that of the total 328.73 m ha geographical area, nearly 188 m ha of land in the country is potentially exposed to various degradation processes. The land area subjected to degradation by way of soil displacement through erosion by water and wind is estimated at 148.9 and 13.5 m ha, respectively (nearly half of the area). About 13.8 m ha is under chemical deterioration due to loss of nutrients and organic matter, salinisation and sodification. Soil acidification also rendered about 49 m ha of land degraded.

Adequate plant nutrient supply holds the key to improving the food grain production and sustaining livelihood. Nutrient management practices have been developed, but in most of the cases farmers are not applying fertilisers at recommended rates. They feel fertilisers are very costly and not affordable and due there is a risk particularly under dry land conditions. Therefore, INM plays an important role which involves integrated use of organic manures, crop residues, green manures, biofertilisers etc. with inorganic fertilisers to supplement part of plant nutrients required by various cropping systems and thereby fulfilling the nutrient gap.

The basic concept underlying the principle of integrated nutrient management is to maintain or adjust plant nutrient supply to achieve a given level of crop production by optimising the benefits from all possible sources of plant nutrients. The basic objectives of IPNS are to reduce the inorganic fertiliser requirement, to restore organic matter in soil, to enhance nutrient use efficiency and to maintain soil quality in terms of physical, chemical and biological properties. Bulky organic manures may not be able to supply adequate amount of nutrients, nevertheless their role becomes important in meeting the above objectives.

Long-term studies being carried out under all Indian Coordinated Research Project have indicated that it is possible to substitute a part of fertiliser N needs of kharif crop by FYM without any adverse effect on the total productivity of the system in major cropping systems such as rice-rice, rice-wheat, maize-wheat, sorghum-wheat, pearl millet-wheat, maize-wheat and rice-maize. Sustainable yield index (SYI) of maize-wheat cropping system after 27 years at Ranchi was the highest with integrated use of 100 percent NPK and FYM. Organic manures alone cannot supply sufficient P for optimum crop growth because of limited availability and low P concentration.

The organic manures are known to decrease P adsorption/fixation and enhance P availability in P-fixing soils. Organic anions formed during the decomposition of organic inputs can compete with P for the same sorption sites and thereby increase P availability in soil and improve utilisation by crops. Reddy *et al.* observed higher apparent P recovery by soybean-wheat system on Vertisol with a combination of fertiliser P and manure.

RECYCLING OF CROP RESIDUES AND GREEN MANURING

Management of crop residues is either through of the following 3 methods; removal, burning or incorporation into soil. Burning is a minor practice in India. Sidhu and Beri reported that in situ recycling of crop residues in rice-wheat rotation reduced grain yield of rice and wheat. Therefore, most of the farmers recycle the crop residues not by choice but due to combine harvesting, burn the residue causing loss of precious organic matter, plant nutrients and environmental pollution. Experiments conducted in Punjab have shown that co-incorporation of green manure and crop residues of wheat and rice helped alleviate the adverse effects of unburned crop residues on crop yields (Table 13).

ROLE OF BIOFERTILISERS IN INM UNDER INTENSIVE SYSTEMS

Several studies clearly indicate that among the different types of biofertilisers available at present, Rhizobium is relatively more effective and widely used. Considering an average N fixation rate of 25 kg N/ha per 500 g application of Rhizobium, it is expected that 1 tonne of Rhizobium inoculants will be equivalent to 50 tonnes of nitrogen. On the other hand, +Azotobacter, which is used in non-legume crops has given inconclusive results. Similarly, Blue Green Algae (BGA) and Azolla have been reported to be effective only in certain traditional rice growing areas in the country.

Table 13. Effects of incorporation of green manure and crop residue on grain yield of rice (t/ha)

Treatment	*1988*	*1989*	*1990*	*1991*	*1992*	*1993*
Control (No N)	4.0	4.6	3.7	4.3	4.1	3.4
150 kg N/ha	6.3	6.6	6.2	6.5	5.7	5.6
180 kg N/ha	6.6	6.9	5.8	6.7	NT	5.3
G.M.	6.6	6.5	6.2	6.5	5.8	5.5
G.M. + wheat straw	6.9	6.9	6.4	6.8	5.6	5.5
G.M. + rice straw	6.9	−6.9	6.7	7.0	5.9	5.3
l.s.d. (P = 0.05)	0.53	0.59	0.46	0.45	0.32	0.37

Meanwhile if BGA applied at 10 kg/ha fixes 20 kg N/ha, then 1 tonne of BGA has an equivalent fertiliser value of 2 tonnes of nitrogen. The beneficial effect of the organisms like Azospirillum and Azotobacter in suppression of soil-borne pathogenic diseases of crops is yet to be established on a pilot scale. Another important role of biofertilisers is liberation of growth substances, which promote germination and plant growth. Against the total anticipated biofertilisers demand of 1 million tonne in the country, the current supply position is very low (<10 000 tonnes). There are several constraints to effectively utilise and popularise the use of biofertilisers. Some of these constraints are:

- Unlike mineral fertilisers, use of the biofertilisers is crop and location specific. A strain found ideal at one location may be ineffective at another location due to competition of native soil microbes, poor aeration, high temperature, soil moisture, acidity, salinity and alkalinity, presence of toxic elements etc;
- Low shelf life of the microorganisms;
- Unlike mineral fertilisers, biofertilisers need careful handling and storage;
- Lack of suitable carrier material, for restoration and longevity in actual field conditions.

In order to overcome the above-cited constraints and make biofertilisers an effective supplementary source of mineral fertilisers, these aspects need to be critically attended.

Constraints in Use of Organics

Convenience and advantages in use of fertilisers

Though fertilisers are costly inputs in agriculture, they are 'concentrated' source of plant nutrients which can be formulated or tailored before or just prior to field application as per needs of the crops and can be applied with minimum transport and labour and at right time. Fertiliser use is high in irrigated crops, commercial crops and in peri-urban areas where awareness is high. The farmers are aware of the need for high nutrient use in high production areas under irrigated condition.

Selective use of fertilisers and manures

Fertiliser use is high in rice, wheat, sugarcane and cotton. Organic manures wherever available are invariably used in some vegetable crops like potato, onion, chillies, spices like ginger and turmic, in cereals like rice, in commercial crops like sugarcane, cotton and fruit crop banana. Green manuring is very prominent in rice and sugarcane and farmyard manure is commonly applied in arid and semi-arid dry land areas where costly fertilisers are discouraged due to the risk associated with their use and also due to the need for water for irrigation/soil moisture for better utilisation of the applied nutrients. Farmers are also aware of the need for organics in dry land agriculture where some sort of stability to production is ensured because of its possible role in soil structure improvement and moisture storage and supply. The farmers in India apply good amount of organic manure (FYM, compost, goat, poultry and pig manure) at some periodicity to regenerate the soil fertility after three to five years of cropping.

Peri-urban/rural differences

Developed market encourages farmers to use fertilisers and produce more under intensive system of cropping. Even small farmers use more fertilisers inputs, however in peri-urban areas, there is also possibility for use of agro-industrial or urban municipal wastes along with fertilisers to augment soil fertility. Farmers' in remote areas with poor infrastructure and without access to market but are aware of the benefits of fertiliser use locally available organic sources.

FARMING SYSTEMS APPROACH

Besides the above stated constraints, to make INM a reality, micro-watershed based agricultural diversification through farming systems approach, consisting of crop and animal husbandry, horticulture, bee keeping, pisci culture, etc. are needed to be adopted. In general terms, the goal of farming systems approach is to increase and stabilise farm production and farm income. Having diverse enterprises creates opportunities for recycling, so that pollution is minimised because a waste in one enterprise becomes an input for another.

The risk minimisation, employment generation and sustained/increased household income are the benefits associated with multi-enterprise farming systems. Appropriate and situation-specific farm diversification models need to be developed and diffused. Efforts are underway in different locations to develop farm diversification models involving judicious enterprise mix that may provide attractive income besides meeting household demands from a given piece of farmland. One such model put forth by Behera and Mahapatra suggests an optimum integration of farm enterprises for a small land holding of 1.25 ha for Bhubaneshwar conditions. In this particular model, land was allocated for different enterprises in proportion to their significance in household needs and demand in local market. It was shown that with the adoption of this diversification model, a net income of Rs.58 360/year was derived from 1.25 ha farm land. This kind of model is worth emulating in parts of the country as well in our search for comprehensive food and nutritional security.

LAND-MANAGEMENT SYSTEMS

Land-management systems have an impact on soil environment. Soil physical, chemical and biological properties are modified considerably by agricultural practices. They may affect soil organisms either positively or negatively, so modifying the size and composition of soil biological communities, with important consequences on soil fertility and plant productivity. Clearing forests or grasslands for cultivation modifies the soil environment drastically and, hence, also modifies the numbers and kinds of soil organisms. The quantity and quality of plant residues and the number of plant species are, in general, greatly reduced, as a result the range of habitats and foods for soil organisms is reduced significantly. The use of large quantities of agrochemicals (pesticides, herbicides and inorganic fertilisers) and tillage practices has a negative influence on soil communities by reducing their numbers and hence the beneficial ecological functions in which they participate.

The beneficial effects of soil organisms on agricultural productivity that may be affected include: organic matter decomposition and soil aggregation; breakdown of toxic compounds, both metabolic by-products of organisms and agrochemicals; inorganic transformations that make available nitrates, sulphates and phosphates as well as essential elements such as iron (Fe) and manganese (Mn); N-fixation into forms usable by higher plants.

Decomposition is the central process in soil. The breakdown of organic residues by soil microflora to release plant nutrients is often accelerated in the presence of soil fauna. The breakdown of plant residues by soil fauna increases the exposure of substrates to the microflora, leading to enhanced nutrient release. In the humid tropics (e.g. Nigeria), the rate of breakdown of plant residues where earthworms and millipedes are present may be twice as high as where they are not present.

Farmers need to create favourable conditions for soil life. They should manage organic matter so as to create a fertile soil in which healthy plants can grow. In tropical agriculture, where poor farmers generally suffer from decreasing soil fertility and declining soil water dynamics, the restoration of SOM is essential for the stabilisation of production. Declining soil water dynamics is partly a result of drought conditions but also significantly affected by loss of vegetation cover, soil crusting and compaction and loss of soil organic matter. These result in reduced surface infiltration, water retention and permeability through the soil as well as increased runoff and hence erosion.

However, this situation cannot be achieved by merely incorporating organic matter into the soil as the degradation process under tropical conditions is too fast to allow any medium- or long-term improvement in soil properties. In addition, the incorporation into soil of organic matter implies tilling the soil, which accelerates its breakdown and destroys soil structure and organisms. The primary need is to feed soil organisms and to regulate their living conditions, while protecting them from harmful chemical and mechanical impacts. For example, shallow tillage, ridge-tillage, or no-tillage and surface management of crop residues has often led to increases in earthworm activity compared with areas where deep tillage is practised. Earthworms are a resource that may be used in agriculture because their effects on nutrient dynamics and the physical structure of soil may significantly enhance plant growth and conserve soil quality. Management options that stimulate the activities of these organisms could promote sustainable production in tropical agro-ecosystems.

For example, the success of earthworm-management techniques may depend on: the choice of suitable species; the provision of adequate organic supplies to feed the worms; and the maintenance of a minimum diversity in all invertebrate communities. Therefore, all these biological resources need to be managed at the same time.

In the humid tropics, "in-soil" technologies that incorporate organic residues into the soil to stimulate the activities of local or inoculated populations of soil-dwelling earthworms should be preferred in most cases to "off-soil" techniques that simply use earthworms to prepare compost. The vermicomposting of residues allows the rapid transformation of fresh residues into compost that can be used readily in the field.

However, a large amount of C is lost that might have been used to sustain mechanical activities of earthworms and other invertebrates in the soil. Endogeic earthworms participate in the humification of organic matter, but they also contribute to: the macroaggregation of soil particles; the maintenance of macroporosity; and the intimate mixing of organic compounds, with effects on the long-term sustainability of soil fertility. Vermicomposting should only be recommended when the quality, the amount or the location of organic residues makes them unsuitable for local use in agriculture. "In-soil" technologies are based on the use of endogeic and anecic earthworms that influence soil physical properties significantly. These technologies manipulate earthworm communities either directly, through the massive inoculation of suitable populations, or indirectly, by promoting suitable conditions for the activity of the already existing population through the manipulation of plant communities and/or organic inputs. The loss of SOM and nutrient deficiency is one of the most important problems facing farmers. SOM is of key importance to optimising crop production, minimising environmental impacts and, thus, improving soil quality and the long-term sustainability of agriculture. SOM benefits crops by providing nutrients, especially N, as it is decomposed by microorganisms. Furthermore, it: (i) binds soil particles together so that they prevent erosion; (ii) improves soil structure and tilth, which allows water, air and nutrients to move readily to living organisms; (iii) is a strong absorber of pesticides, organic wastes and heavy metals; and (iv) is a part of the cation exchange complex, which holds many nutrients and resists losses to leaching.

In a natural grassland or forest ecosystem, SOM accumulates with soil development and eventually reaches an equilibrium, which is determined mainly by the environment, natural vegetation and soil organisms. A decrease in SOM leads to a decline in soil aggregate stability and in crop yield. Where soil is tilled and cropped, massive changes occur within the soil system.

Tillage mixes oxygen into the soil and breaks up its structure, giving microbes all they need in order to burn up organic matter and release CO_2 into the atmosphere more quickly than when this process is realised naturally by soil organisms. In the natural state, much of the SOM is protected. The soil system can be thought of as a structure with millions of tiny pores of all sizes and shapes. Much of the SOM is trapped in areas inaccessible to microbes. Other organic particles are clumped (aggregated) together, so allowing microbes access to the outside while the centre remains protected. Soil organisms such as earthworms mix SOM and mineral particles, thereby facilitating its access by microorganisms. Tillage speeds up this process; stirring, churning and mixing everything together. Microbes suddenly find a feast of nutrient rich material and proceed to multiply, liberating excess nutrients that, if not taken up by the crop, vulnerable to be leached out of the soil with rain or irrigation water.

The rate and degree of organic matter depletion is influenced by: (i) type of crop esidues; (ii) tillage practices; and (iii) type and severity of wind and water erosion. In conventional agro-ecosystems, crop residues are removed or burned after harvesting. Jnless other organic materials are added to the soil, this will lower the amount of SOM, and so inorganic nutrient inputs are needed to sustain plant growth. The loss of SOM ias a negative effect on soil macroinvertebrates, whose abundance decreases as a consequence of the reduced amount of food available. In addition, conventional systems are characterised by repetitive tillage, which physically disturbs the soil and reduces greatly the abundance of soil macrofauna. Most SOM is contained in the surface horizons of soils. Soils become more sensitive to erosion under such management practices, and the physical removal of topsoil by wind and water erosion can result in further significant loss of organic matter.

Organic matter losses can be minimised through the application of conservation arming practices, which include: using permanent crop covers to prevent erosion and dd organic matter; employing crop rotations that include forages and legumes; adding nanures and organic wastes to supplement crop residues; realising mulching on the oil surface; using an adequate and balanced fertiliser programme that creates healthy rops and good residues; using no-tillage where possible or minimum tillage; in no-illage, seed is drilled directly into the stubble or the surface residues which have been lattened with a knife roller or killed with a herbicide without first ploughing the soil.

The use of no-tillage in different kinds of crops in the Parana region (southern Brazil) as improved soil environmental conditions for plants and soil animals compared with oils under conventional tillage management systems. Results include: reduced erosion; nhanced nutrient-use and water-use efficiency by crops; and improved crop yields and rofitability, especially after a transition period of a few years. No-tillage practices have lso increased soil macrofauna diversity and accelerated population recovery after the essation of conventional tillage practices. Soil organisms that have benefited especially om no-tillage are: natural predators; bioturbators; and decomposers. Finally, the lack f soil disturbance at no-tillage sites has led to: increased SOM in the top-most soil layers; creased protection of the soil surface with plant residues; and increased populations f beneficial soil invertebrates.

In contrast to ploughed systems, no-tillage management leads to accumulation of lant residues on the soil surface. This decreases the rate of decay of crop material and, erefore, helps to maintain good SOM levels. The adoption of soil conservation ractices that reduce soil erosion will assist in reducing losses and maintaining SOM.

Increasing the amount of crop residue will also assist in maintaining and/ or creasing SOM levels. This can be accomplished by selecting crops that produce more residue, by adequate fertilisation and by rotations of cereals with legumes and forage crops, especially those with deep extensive rooting, as well as minimising the removal or burning of crop residues. In sandy soils or soils low in organic matter, the addition of manure and the inclusion of legumes in the rotation, or as a green manure crop, will be needed to enhance SOM and minimise fertiliser losses (if used) through leaching.

Conventional tillage practices, based on the use of hand hoes, ploughs – animal drawn and powered – and harrows are likely to destroy soil structure and make the soil vulnerable to compaction and erosion. Wheel traffic or pressure exerted on the soil surface by large animals, vehicles and people can cause soil compaction. Compaction occurs where moist or wet soil aggregates are pressed together and the pore space between them is reduced. Compaction changes soil structure, reduces the size and continuity of pores, and increases soil density. Compaction reduces the capacity of the soil to hold water, the rate of water movement through the soil, and the storage capacity of the soil. Compaction and crusting of the soil surface also limits water infiltration resulting in increased runoff and vulnerability to erosion and hence further loss of potential productivity. When the amount of water that enters the soil is reduced, less water is available for plant growth and percolation to deep root zones. Soil compaction can also be caused indirectly by a decrease in soil organisms and hence loss of biological tillage.

Deep tillage is harmful to soil organisms. It can kill them outright, disrupt their burrows, lower soil moisture, and reduce the amount and availability of their food. Other inappropriate land-management practices, such as the use of certain pesticides and some inorganic fertilisers, can also be harmful to soil life. All these practices result in declining soil life and SOM, which are important for oxygen, water and nutrient cycles, including moisture retention, water infiltration and plant nutrition. In general, soil tillage reduces the abundance of soil organisms. Termites, earthworms, beetles and spiders are among the main groups of soil macrofauna usually much reduced by tillage practices. As they are reduced, their functions in soils are altered, leading to soil degradation and increases in pests and plant diseases.

Compaction reduces the capacity of the soil to hold water, the rate of water movement through the soil, and the storage capacity of the soil. Compaction and crusting of the soil surface also limits water infiltration resulting in increased runoff and vulnerability to erosion and hence further loss of potential productivity. When the amount of water that enters the soil is reduced, less water is available for plant growth and percolation to deep root zones.

The alteration of soil macrofauna abundance and diversity as consequence of soil-management practices may lead to serious problems of soil compaction. The functions previously performed by depleted organisms are no longer performed and the activity of surviving organisms may become excessive and produce a negative effect on soil. In the Brazilian Amazon (Manaus) the transformation of the forest zones into pastures, the use of agricultural machinery and trampling by cattle has led to severe soil compaction, particularly in the surface layer (5–10 cm). However, the most important consequence was that the native soil macrofauna communities were radically altered, most of the native taxa disappearing. The activity of the compacting earthworm species Pontoscolex corethrurus produces more than 100 tonnes/ha of castings. The excessive accumulation of these compacted structures is wholly prevented by small invertebrates that feed on them and break them down into smaller structures. However, where these organisms are not present because of soil compaction by cattle or machinery, this function is not performed and the accumulation of compacted casts affects plant growth negatively.

Where soil density increases significantly, it limits plant growth by physically restricting root growth. Severe compaction can limit roots to the upper soil layers, effectively cutting off access to the water and nutrients stored deeper in the soil. Anaerobic conditions (lack of oxygen) can develop in and above the compacted layer during wet periods, further limiting root growth. Compaction alters soil moisture and temperature, which control microbial activity in the soil and the release of nutrients to plants. Anaerobic conditions increase the loss of soil N through microbial activity. Compaction also changes the depth and pattern of root growth. These changes affect the contribution of root biomass to SOM and nutrients. By reducing the number of large pores, compaction can restrict the habitat for the larger soil organisms that play a role in nutrient cycling and can thus reduce their numbers.

Biological activity may be used to reduce soil compaction. In Burkina Faso, organic mulch (cow manure or straw) was applied to soil surfaces in a three-year study in order to trigger termite activity. The termites restored crusted soils through their burrowing and decomposing activities, properly managed by careful additions of organic matter. The increase in soil porosity and the improvement in water infiltration and retention capabilities of the soil encouraged root penetration and hence crop productivity. Termites can become a major crop pest in drylands when there is inadequate alternative dead and dried up organic matter – their preferred food source. Best practices and deterrents need to be researched and adapted with farmers for specific farming systems. For example, making use of termite predators, such as ants and birds, through diversified farming systems, reducing plant water stress and vulnerability to termite attack, and use of plants with repellent properties.

Soils with a well-developed structure and high aggregate stability are more resistant to compression than other soils. Near-surface roots, plant litter, and aboveground plant parts reduce the susceptibility to compaction by helping to cushion impacts. Vegetation also adds SOM, which strengthens the soil, making it more resistant to compaction. Conventional tillage, especially in tropical latitudes, causes rapid oxidation and mineralisation of organic matter and, as a result, declining SOM levels.

Conservation tillage is linked with increased earthworm activity; mulched stubble in particular favouring large increases in earthworm numbers. The retention of maximum levels of crop residues on the soil surface and a lack of soil disturbance creates a more favourable habitat for soil animals. Earthworm channels and termite galleries increase the volume of soil pores, which should increase soil aeration and water entry into the soil. The number of large soil pores under no-tillage practices can be up to sevenfold greater than under conventional tillage systems. Another practice to improve a compacted soil is the growing of plants with large taproots that are more effective at penetrating and loosening deep compacted layers, and the use of shallow, fibrous root system to break up compacted layers near the surface. Roots also reduce compaction by providing food that increases the activity of soil organisms.

Conventional tillage methods are a major cause of severe soil loss and desertification through soil compaction and accelerated erosion by wind and water. Worldwide, it is responsible for about 40 percent of land degradation. Agricultural lands could be protected or saved from degradation and erosion by applying an environmentally friendly tillage approach and the use of cover crops and mulching practices.

The intensity of tillage is a recognised factor affecting the amount of runoff and soil erosion that occur in croplands. In turn, the amount of runoff and soil loss directly affect soluble and absorbed chemical transport, particularly plant nutrients from fertiliser application. Soils with no-tillage management practices present a much lower sediment loss than soils under conventional practices. No-tillage keeps crop residues on the soil surface, which is highly effective in controlling the loss of sediment from water-runoff events. The loss of soil nutrients such as soluble N and P is reduced by up to 50 percent where no-tillage is adopted.

Conventional tillage, especially in tropical latitudes, causes rapid oxidation and mineralisation of organic matter and, as a result, declining SOM levels. Thus, no tillage practices are highly preferable as they leave a protective blanket of leaves, stems and stalks from the previous crop on the surface, to protect the soil from erosion, minimise compaction and maintain soil structure, enhance infiltration and provide organic matter for soil biological activity. These also lead to enhanced fertiliser efficiency.

There are several ways of modifying tillage systems to improve soil resistance to erosion and enhance soil macrofauna populations: leave a rough surface or adequate residue cover for erosion control; bury crop residues for pest control, depending on the system; leave as mulch cover when possible; do as little physical disruption (tillage) as possible; use crop rotations with specialised cover crops – essential for breaking the cycles of weeds and pests and for compensating for higher-value crops with inadequate quantity and quality of residues, such as edible beans and soybeans, with high C:N ratios in their straw.

The adoption of conservation tillage is sometimes restricted by concerns about potential increases in certain pests, particularly plant pathogens and insect pests. The complex interactions within a more diverse soil community in no-tillage may provide protection against the pest organisms, but the time required for this shift in communities may be too long for farmers' economic requirements. The use of synthetic chemical pesticides, particularly herbicides, may be unavoidable in the early years, but they have to be used with care in order to reduce the negative impacts on soil life. To the extent that a new balance between the organisms of the farm-ecosystem, pests and beneficial organisms, crops and weeds, becomes established and the farmer learns to manage the cropping system, the use of synthetic pesticides and mineral fertiliser tends to decline to a level below the original "conventional" farming level.

Where soil insects become a pest, they contribute to a poor yield. The damage caused by pests can be responsible for 40–50 percent of crop loss. Groundnut plants are particularly susceptible to attacks by soil insects because the pods develop underground. Groundnut culture is an important crop and component of the diets of many smallholders in sub-Sahelian Africa. They have access only to 1–2 ha, from which they derive the food to support their family and sometimes all their income.

Soil organisms like white grubs, e.g. Diabrotica larvae, some species of termites, millipedes, wireworms (Elateridae), false wireworms (Tenebrionidae) and ants can became serious groundnut pests in Africa, Asia, and Central and South America. All these organisms are podborers and attack roots, causing important crop loss.

There are four basic approaches for managing soil insects in groundnut crops: apply insecticides, alone or in various combinations; manipulate natural control processes; rely on host plant resistance; employ management practices that prevent the insects from multiplying unduly The application of insecticides is not recommended because they are usually not sufficiently effective and they are expensive. Aside from the well-known hazards that these chemicals present to the environment and humans, they can be accumulated by plants. These chemicals are not pest specific and kill other beneficial organisms present in soils. This can led to the proliferation of other organisms that become more harmful. For example, in a Malawi field, there was a proliferation of termites following an insecticide application because of the disappearance of their natural enemies, the ants. It is most important that the natural control processes should not be disturbed. Thus, the unavailability of insecticides in many African countries is not necessarily a bad thing. The abundance of natural enemies indicates that there is little reason for considering the release of exotic or any other natural enemies. However, any management procedure that encourages the proliferation of suitable predators and parasites should be considered the best solution.

In Nigeria, a way of controlling termite invasions is to cultivate plants that have repellent or antibiotic properties, or at least not to remove them from the fields. The plants include: basil, termite grass Vetiveria nigritana, Digitaria sp., lemon grass Cymbopogon shoenanthus, and elephant grass Pennisetum purpureum. Some farmers also introduce soldier ants into termite mounds. Spacing can also have a significant effect on the damage caused by pests. A study in India showed that in plots with 15-cm spacing between cotton plants in rows spaced 75 cm apart, the average percentage of attacks by the ash weevil grubs was 17 compared with 22 percent in 20-cm spacing and 31 percent in 30-cm row spacing.

Crop rotation and intercropping using plants that are not hosts for a certain pest are very important to preventing the survival and the continuous reproduction of the pest throughout the year.The best solution to problems caused by pests could be a whole-farm ecological approach to pest management that considers beneficial organisms as mini-livestock that can be managed. These organisms could develop more readily and be more effective biocontrols when provided with a habitat with adequate and easily available resources. This way to increase the habitat for beneficial organisms should be understood and practised within the context of overall farm management goals.

REFERENCES

Bollen, Walter B. 1959. *Microorganisms and Soil Fertility*. Oregon State College. Oregon State Monographs, Studies in Bacteriology, No. 1. 22 p.

Boyle, M., W.T. Frankenberger, Jr., and L.H. Stolzy. 1989. "The influence of organic matter on soil aggregation and water infiltration". *Journal of Production Agriculture*. Vol. 2. p. 209-299.

Comis, Don. 1997. "Glomalin—soil's superglue". *Agricultural Research*. USDA-ARS. October. p. 22.

Jackson, William R. 1993. *Organic Soil Conditioning*. Jackson Research Center, Evergreen, CO. 957 p.

Kumar, P. 1998. "Food demand and supply position for India", *Agric. Econ. Policy Paper* 98-01, IARI, New Delhi.

Sehgal, J.L. and Abrol, I.P. 1994. *Land Degradation in India: Status and Impact*. Oxford & IBH Publishing Co. Pvt. Ltd., New Delhi.

Sekhon, G.S. 1997. In: *Plant nutrient needs, supply, efficiency and policy issues*: 2000-2025 (J.S. Kanwar and J.C. Katyal eds.), NAAS, New Delhi, pp. 78-90.

Tandon, H.L.S. 1995. In: Proceedings of the IFPRI/FAO workshop on plant nutrient management, food security, and sustainable agriculture: The future through 2020 (Eds.: P. Gruhn, F. Goletti and R.N. Roy), Viterbo, Italy, May 16-17, 1995. pp. 199-234.

9

Protecting Plantations from Diseases

Damaging pests and diseases are found in plantations and natural forests in both the temperate and tropical zones. A fundamental concept of ecosystem dynamics is that as diversity increases, so does stability. The greater the number of plants and animals that occupy an ecosystem, the greater are the checks and balances that prevent any one species from increasing to the point where other ecosystem components are threatened. With pests and diseases, the greatest single deterrent to population increase is the amount of available host material. For example, in a complex tropical rain forest, where as many as a thousand plant species might occupy a single hectare, a population of a host specific, phytophagus caterpillar will find only a limited amount of suitable host material. Therefore its numbers will remain fairly stable.

On the other end of the diversity scale is the agro-ecosystem. Dominated by a single plant species, agro-ecosystems often cover large areas and provide a virtually unlimited amount of suitable host biomass. This can result in population explosions of organisms that use the plant as host material. Furthermore, the lack of ecosystem diversity will, at best, provide only a marginal habitat for the natural enemies of the organisms that are using the crop plant as host material. In terms of diversity and stability, forest plantations tend to be more like agro-ecosystems than natural forests. This would lead one to conclude that plantation forests are more susceptible to damage by pests and disease than are natural forests. While this may be generally true, it does not necessarily mean that natural forests are immune to damage from these agents. The relative lack of pest outbreaks in mixed tropical forests is often cited as evidence for the importance of diversity in stabilising plant communities.

Many natural forests in the northern hemisphere boreal and temperate zones however, tend to be relatively simple ecosystems when compared to tropical forests and are susceptible to outbreaks of insects. Examples include:

- the spruce budworm, *Choristoneura fumiferana,* a defoliator of *Abies balsamea* and *Picea* spp in the boreal conifer forests of eastern Canada and adjoining states in the USA,
- western spruce budworm, *Choristoneura occidentalis,* and Douglas-fir tussock moth, *Orgyia pseudotsugata,* in natural conifer forests of western North America,
- the processionary caterpillar, *Thaumetopoea pityocampa,* in Mediterranean Europe and North Africa
- and several species of conifer attacking bark beetles, *Dendroctonus* spp and *Ips* spp in the forests of Canada, the United States, Mexico, Central America and the Himalayas.

Natural forests are also subject to damage by a variety of diseases caused by fungi, bacteria, virus or parasitic mistletoes and they are vulnerable to introductions of exotic pests and disease. The introduction of white pine blister rust, *Cronartium ribicola,* into North America, the European gypsy moth, *Lymantria dispar,* into the mixed broadleaf forests of the eastern United States, of two *Juniperus* infesting scale insects, *Carulaspis visci* and *Lepidosaphis newsteadi,* into Bermuda and the apparent introduction of the soil fungus *Phytophthora cinnamomi* into *Eucalyptus marginata* forests in Western Australia are just a few examples. Moreover, the recent introduction of pitch canker disease, caused by the fungus *Fusarium subglutinans* f. sp. *pini,* into forests of *Pinus radiata,* along the California, USA coast, now is a source of concern.

A more recent example is the discovery in 1994-5 of the pink or hibiscus mealybug, *Maconellicoccus hirsutus,* an insect capable of feeding on between 125 and 150 different plant species, in the Caribbean islands of Grenada, Trinidad and Tobago and St. Kitts-Nevis. This insect poses a severe threat to natural mixed tropical forests as well as vegetable crops and ornamental plants in the region.

EXOTIC SPECIES PLANTATIONS

A typical scenario in the use of exotics in forest plantation programmes, both in tropical and temperate zones, is that they enjoy a pest free "honeymoon" for a period of time. During this period, management practices often evolve that make the plantations especially susceptible to pests and disease once they begin to appear. These include:

1. Failure to give proper attention to species/site matching.
2. Use of planting stock from a narrow genetic base.
3. Failure to maintain optimum stocking levels and tree vigor through intermediate cuttings.

4. Dependence on one or two species in a plantation programme, resulting in an unlimited supply of host material for potentially damaging organisms.

When pests and diseases do begin to appear, they may come from two sources. A complex of indigenous agents may adapt to the new host or exotic pests may be accidentally introduced.

Indigenous Pests and Diseases

A number of examples of indigenous insects adapting to exotic forest plantations are known. In Colombia, indigenous insects have been detected in that country's exotic *Cupressus, Eucalyptus* and *Pinus* plantations, with the most frequently encountered groups being defoliating caterpillars of the family Geometridae, leaf cutting ants and stem infesting insects. Rodas P. reports 30 species of defoliating insects feeding on exotic forest plantations in the Andean region of Colombia.

A similar situation exists in Chile, where several species of indigenous pests are now associated with exotic forest plantations including *Ormiscodes cinnamomea,* a large Saturniid that occasionally defoliates *Pinus radiata*. Ohmart and Edwards report 96 indigenous insects feeding on various species of *Eucalyptus* in China, 94 in India, 223 in Brazil, 31 in New Zealand, 105 in Papua New Guinea and 62 on the island of Sumatra, Indonesia. Termites and leaf cutting ants are the major damaging agents.

Indigenous disease causing organisms are also capable of adapting to new hosts and causing severe damage. In Kenya, plantations of *Cupressus macrocarpa* have been severely damaged by *Monochaetia unicornis,* a stem canker causing fungus. During the early 1990s, several species of foliar and twig pathogens, including Cylindocladium leaf blight, Coniella leaf spot, Kirramyces leaf spot and pink disease, *Corticum salmonicolor*, damaged *Eucalyptus* plantings in Vietnam.

Exotic Pests and Diseases

The greatest hazard of pest and disease damage to exotic forest plantations is from the accidental introduction of exotic pests and diseases. Exotic pests and diseases agents, in the absence of natural enemies, and provided with a large supply of suitable host material, that may have little resistance to the new pest or disease, can build up rapidly and cause devastating losses.

One of the best-documented examples of exotic pests and diseases causing damage to exotic plantations is to pine plantations in the Southern Hemisphere, especially *Pinus radiata. P. radiata* were first planted in the late nineteenth century in Australia, Chile, New Zealand and South Africa. The excellent growth and form of the tree in these locations encouraged additional plantings. These provided the basis for thriving lumber and paper industries, especially in Chile and New Zealand. One of the first exotic insects to appear in New Zealand's radiata pine plantations was the European wood wasp, *Sirex noctilio,* during the early 1900's. This insect was probably introduced on unprocessed pine logs imported from Europe.

Native to southern Europe and North Africa, where it causes little or no damage, *S. noctilio* normally breeds in recently killed pines. Plantations established in New Zealand during 1920-30 stagnated because there was no market for small logs from thinning operations. This made the plantations susceptible to *S. noctilio* and its associated fungus, *Amylostereum areolatum*. By 1947, high levels of tree mortality were occurring, primarily in the unthinned plantations. However, today *S. noctilio* is not a problem because of biological control methods and better stand management.

S. noctilio subsequently appeared in *P. radiata* plantations in Australia in 1952. In 1980 infestations were detected in South America where this insect is damaging plantations of *P. elliotii* and *P. taeda* in Argentina, southern Brazil and Uruguay. In 1994, infestations of *S. noctilio* were discovered in *P. radiata* plantations in the Cape Peninsula of South Africa.

Other damaging insects that have been introduced into *P. radiata* plantations include the bark beetle, *Ips grandicollis,* a North American species, into Australia and the European pine shoot moth, *Rhyacionia bouliana* into Argentina and Chile. The latter is now regarded as Chile's number one forest insect pest. The European bark beetles, *Hyalstes ater* and *Hylurgus ligniperda* were introduced into Chilean radiata pine plantations during the late 1980s and are capable of killing young pine seedlings.

Several introduced fungi have also caused severe damage to *P. radiata* plantations. In 1957, a needle fungus, *Dothistroma septospora* (*D. pini*), became a major concern in *P. radiata* plantations in Chile, parts of East Africa and New Zealand, although there are records of its occurrence in these areas as early as 1940. The stem disease, *Sphaeropsis sapinea* (*Diplodia pini*) has also been introduced in a number of countries. This fungus has been especially damaging in Kenya, where it first appeared in 1973 and proliferated on *P. radiata,* a species adapted to a winter rainfall regime and is especially susceptible to fungal attacks during periods of rain and warm temperatures.

Another example of damage to exotic forest plantations by an introduced pest is that of the leucaena psyllid, *Heteropsylla cubana* on *Leucaena leucocephala,* a fast growing legume native to Mexico and Central America. This tree was introduced into the Philippines as early as 1565 and was quickly established throughout the tropics where local people found it an excellent fuelwood and later as shade under which crops such as coffee, cocoa, pepper and vanilla could be grown. Genetically improved varieties were eventually introduced into the tropics where it was regarded as having the widest range of uses of all of the tropical legumes. In 1983, leucaena psyllid, a somewhat obscure insect of *L. leucocephala* in its natural range, was detected in southern Florida, USA and during the following year it was found in the Hawaiian Islands. The insect spread rapidly across the Asia and Pacific Region between 1985 and 1988. In 1991, infestations were found on the southern Indian Ocean islands of Reunion and Mauritius and one year later in several eastern African countries, including Burundi, Kenya, Tanzania and Uganda. This insect feeds on the shoots and young foliage of Leucaena causing wilting and growth loss.

Ironically, attempts to introduce new genetic material into a country have occasionally resulted in introductions of damaging pests. In 1968, a pine woolly aphid, *Pineus borneri,* was introduced into Kenya on *Pinus caribaea* grafts imported from Australia. Twenty years later, a scale insect, *Oracella acuta,* was introduced into China on *P. elliotii* scion material introduced from the United States. By 1995, over 212,500 ha of *P. elliotii* plantations in Guangdong Province in southern China were infested by this insect.

INDIGENOUS SPECIES PLANTATIONS

Throughout much of Asia, Europe and North America, forest plantations are typically established with indigenous species. These plantations are also susceptible to damage by a complex of both indigenous and exotic pests and disease.

Indigenous Pests and Diseases

In China and Vietnam, the defoliating caterpillar, *Dendrolimus punctatus,* is a major pest of indigenous pine plantations. This insect has two generations per year in east-central China where it defoliates *Pinus massoniana*. In Vietnam, *D. punctatus* has four generations per year and outbreaks have occurred in plantations of *P. merkusii,* often established on steep deforested slopes where they are an important source of lumber, pulpwood, fuel wood and resin.

Nun moth, *Lymantria monacha,* is a defoliating caterpillar of both *Picea abies* and *Pinus sylvestris* in central Europe. This insect was originally a major pest of natural *Picea abies* forests but readily adapted to the extensive pine plantations. In Poland, plantations of *P. sylvestris* are established on low nutrient, sandy soils or abandoned marginal agricultural lands. Between 1978 and 1983, a major outbreak of this caterpillar necessitated aerial spraying of 2.5 million ha, over 25% of Poland's forests.

In 1991, this insect again reached outbreak proportions and large areas of forest were treated between 1992 and 1994. Other defoliating insects of *P. sylvestris* plantations in

Europe include *Bupalis piniarius, Panolis flammea, Dendrolimus pini* and sawflies, *Diprion* spp.

Fusiform rust, caused by the fungus *Cronartium quercuum* f.sp. *fusiform,* has become a serious pest of *P. elliottii* and *P. taeda* plantations in the southeastern USA. Prior to 1900, the disease was rare, but because of intensive cultural practices associated with plantation forestry, including planting of infected seedlings, fire management, genetic selection of fast growing trees and expansion of the range of susceptible hosts, this disease has become a major problem. Recent estimates indicate that $US130 million in damage is occurring annually and the disease is increasing at the rate of 2-3% per year.

In China, species of *Paulownia,* are widely planted either as ornamentals in villages or in intercropping systems. Pawlonias are commonly infected by a mycoplasma-like organism (MLO), which produces large witches brooms in tree crowns. This disease causes significant reductions in increment, leading to a gradual decline and death of infected trees. The wood of infected trees is of reduced quality.

Exotic Pests and Diseases

The pine wood nematode, *Bursaphelenchus xylophilus* is the cause of pine wilt disease. This disease has been a major cause of tree mortality in pine plantations in Japan since the early 1900's. The nematode is probably native to North America and has been introduced into Japan and more recently China, where it is causing extensive tree mortality in plantations of *Pinus massoniana.* The nematode is spread from tree to tree by longhorn wood boring beetles, *Monochamus* spp, when the adult beetles feed on pine shoots.

MIXED VS SINGLE SPECIES PLANTATIONS

The concept of diversity leading to ecosystem stability would suggest that mixed species plantations would be more pest and disease resistant than single species plantation. Indeed, work by Klimetzek in Germany indicates that the inclusion of broadleaf trees in pine forests, either as a mixture or in the understory, tends to reduce the incidence of outbreaks of *Lymantria monacha.* On the other hand, planting *Picea abies* under a plantation of *P. sylvestris,* could intensify or prolong outbreaks of this insect since both trees are hosts.

Wormald indicates that data on relative susceptibility of mixed and pure species plantations in subtropical and tropical regions is, at best, inclusive and confusing. While examples of where pests and diseases have become problems in pure species plantations are cited, many result from accidental introductions or of indigenous pest species adapting to new conditions. Furthermore he cites examples where establishment of mixed species plantations has had little or no effect on reducing pest caused losses. Some temperate zone pests and diseases have adapted to mixed-species ecosystems by having evolved life stages on alternate hosts.

Many rust diseases fall into this category, including fusiform rust of *P. elliotii* and *P. taeda, Cronartium quercuum* f.sp. *fusiform,* which has alternate spore stages on oak, *Quercus* spp. Therefore, removal of oaks from the vicinity of pine nurseries in the southeastern USA is a common although not altogether effective practice. Cooley spruce gall aphid, *Adelges cooleyi,* an insect indigenous to western North America, has alternating life stages on *Picea* and *Pseudotsuga menziesii,* trees that are often found growing in close proximity to one another in natural forests.

Perhaps a greater concern with regard to pests and diseases is not mixed vs. single species plantations, but the number of species used in a country's plantation programme. Reliance on one or two closely related species may meet a country's needs for wood products and foreign exchange. However, should a pest or complex of pests appear in these plantations, the results could be devastating. Yet, many countries have fallen into this dilemma.

Kenya, where 46% of its industrial plantation area is composed of *Cupressus lusitanica;* Colombia with 66% of its plantations *Pinus patula;* Poland with 70% of its plantation forests *P. sylvestris;* and Chile and New Zealand with 90% of their plantations composed of *P. radiata.* New Zealand experience has been that deliberately planting a wide range of species to reduce risk from pests and diseases did not work. Rather, good site matching and silviculture, coupled with emphasis on disease prevention and control, backed by research, has been used as an alternative strategy.

MANAGING PESTS AND DISEASES

The objective of managing pests and diseases in forests and forest plantations should be to keep them in a healthy, productive condition. What is a healthy forest? The concept of a healthy forest became a popular concept during the mid- to late 1980s. In economic terms, a "healthy" forest can be defined as "a forest in which pests and diseases remain at low levels and do not interfere with management objectives." In ecological terms, a healthy forest is "a fully functional ecosystem; one in which all of its parts can interact in a mutually beneficial way."

The healthy forest concept directs forest managers to focus on the forest rather than its pests and diseases and takes into account the natural role of insects, fungi, fire and other so called "damaging agents" and their interactions in forest dynamics. Under an overall policy of forest health protection, pests and diseases are looked upon as a symptom of an unhealthy forest rather than as the problem. This directs forest managers and forest protection specialists to address the underlying causes of the pest or disease - factors such as overstocking, over-maturity, poor site/species matching, excessive fuels and single species forests with little diversity. Striving for healthy forests also involves anticipating pests and diseases based on historical records of their occurrence and the knowledge of forest and climatic conditions that favour their abundance. This allows time to implement management practices that will make these forests inhospitable for build-up of damaging pests and disease.

The healthy forest concept may sound like an idyllic approach to managing forest pests and diseases, especially in cases where an introduced pest is causing widespread damage in a simple plantation ecosystem in the absence of its normal complex of natural enemies. Yet even the most damaging of exotic pests and diseases respond to stand and site conditions and there is evidence that they focus their activities on "unhealthy" forests. A classic example is that of *Sirex noctilio* in exotic pine plantations in the Southern Hemisphere.

In New Zealand, it was noted that stressed trees in unthinned plantations were the first to be attacked by this insect. Therefore, by its action, the plantations were effectively thinned. Thinning is now an integral part of managing plantations threatened by *S. noctilio* in places where this insect has been introduced. In another example, when a cypress aphid, originally identified as *Cinara cupressi*, appeared in eastern and southern Africa, most severe damage occurred in plantations of *Cupressus lustianica* that were either overmature or had been established on low nutrient soils.

INTEGRATED PEST MANAGEMENT

The healthy forest concept provides the umbrella or objective for addressing damage caused by pests and diseases. Integrated pest management (IPM), on the other hand, provides the tools to accomplish this task.

The concept or philosophy of IPM as a "rational" approach to pest control was formalised during the 1960s. Crop protection specialists had become aware of the adverse side effects of dependence on chemical pesticides, including pesticide resistance, occurrence of secondary pests, environmental damage and human health hazards. This led to the realisation that alternative approaches, including cultural, biological and genetic tactics, used either alone or in combination, were also needed to provide long-term, effective protection against damaging pests.

Numerous definitions of IPM appear in the literature. Smith *et al.* refer to IPM as "a process based on ecological principles and integrates multi-disciplinary methodologies in developing agro-ecosystem management strategies that are practical and effective and protect both public health and the environment." Pimentel describes IPM as a "pest control method that includes judicious use of pesticide and non-chemical technologies - all of which are based on sound ecological principles." IPM can be looked upon as consisting of two basic elements; a decision process and an action process. The decision process establishes the basis for any subsequent actions to be undertaken, including no action. The action program may consist of one or more ecologically, economically and socially acceptable tactics designed to reduce pest populations to non-damaging levels.

The Decision Process

The decision process is often the most time consuming and complex aspect of IPM. It requires careful consideration of the pest, its host, resource management objectives and the ecological, economic and social consequences of the various available tactics. Population levels of pests are estimated and anticipated resource losses are projected, as are the costs of treatment and its anticipated benefits. If treatment costs exceed losses, a rational decision may be to not treat and accept the losses. Other questions to address include: will natural controls take over within a short enough time so that artificial controls will be unnecessary; or will the effects of proposed treatments be so adverse that they would outweigh the benefits of treatment?

Monitoring of forest pests and diseases and their resultant damage is a critical input to the IPM decision process. Pest monitoring is becoming a sophisticated process that makes use of many technologies. Pheromones and other chemical regulators are often used to monitor insect population levels. Remote sensing technologies such as aerial sketch-mapping, aerial photography and airborne video are used to map and assess forest damage. Geographic information systems (GIS) can be used to relate the location of affected areas to key resource values, terrain features, land ownerships and environmentally sensitive areas. Mathematical models can predict resultant damage caused by certain levels of pest numbers and their consequences. In some cases, pest, growth and yield and economic models are linked to make projections of pest and disease impacts.

The Action Process

IPM action programmes consist of two overall strategies: prevention or direct control (suppression). Specific pest management tactics exist under each of these strategies.

Prevention

Prevention consists of tactics designed to either reduce the probability of the occurrence of a pest or disease or to create environmental conditions inhospitable for its buildup into damaging numbers. Regulatory, cultural or genetic tactics are examples of prevention strategies.

Regulatory tactics are designed to prevent introductions of exotic pests and diseases and to prevent their spread once established. Examples include inspection of wood products and wooden containers at ports of entry to intercept pest species, conduct of pest risk analyses when new trade agreements are made and establishment of quarantine zones when a pest species is first discovered in a new location.

Cultural tactics are designed to create conditions inhospitable for the development of damaging numbers of pests and diseases. These include matching tree species selected for planting to suitable growing sites, controlling stocking through intermediate harvests to maintain tree vigor and timely harvesting of plantations when they reach maturity.

A drastic but sometimes necessary cultural approach is to simply eliminate a tree species from a plantation programme because of its high susceptibility to certain pests and diseases. An example is Kenya's policy to discontinue planting of *Cupressus macrocarpa* due to the stem canker fungus, *Monochaetia unicornis*, and *P. radiata* because of its susceptibility to *Sphaeropsis sapinea*.

Genetic tactics make use of varieties of host plants that are either more tolerant to damage or less palatable to the pest. Identification and testing of varieties of *Leucaena leucocephala* and hybrids with other species of *Leucaena* for resistance or tolerance to the psyllid, *Heteropsylla cubana*, was a major line of investigation following this insect's introduction into the Asia-Pacific region. In the southeastern United States, screening and development of pine seedlings resistant to fusiform rust is an integral part of the management of this disease. Tree breeding programmes in Brazil address resistance to the canker *Cryphonectria cubensis* on eucalypts and breeding of *P. radiata* in New Zealand includes a rating for resistance to the needle fungus *Dothistroma pini*.

Suppression

Tactics directed against the pest or disease are referred to as direct control or suppression tactics. Examples include various types of biological, mechanical or chemical methods.

Biological control involves the use of natural enemies of a pest or disease to help keep its numbers in check. Classic biological control, a technique widely used in agriculture, involves the importation of natural enemies to help control an exotic pest. In Colombia, the geometrid *Oxydia trychiata*, an indigenous species attacking pines and cypress, was successfully controlled by the introduction and release of an egg parasitoid *Telenomus alsophilae*.

In another successful example of classic biological control, releases of a parasitoid, *Pauesia bicolor*, resulted in the collapse of an aphid, *Cinara cronartii*, a North American species accidentally introduced into pine plantations in South Africa. In Chile, state-of-the-art mass rearing facilities have been established by forest industry for production of the European pine shoot moth parasitoid, *Orgilis obscurator*. One such laboratory, *Controladora de Plagas Forestales*, was established in 1992. In 1996, this facility produced 1.2 million parasitised larvae and 22,600 adult female parasitoids for field release.

A key concern about biological control is the possibility that the introduced natural enemy might also attack innocuous or beneficial insects in the ecosystem. Therefore it is necessary to thoroughly evaluate candidate species prior to release to ensure their relative host specificity. Another concern is the hazard of accidentally introducing hyperparasites, natural enemies of the biological control agents, which might eventually affect the agent's efficacy.

Colonies of the leaucana psyllid predator *Olla v-nigrum*, released on the Indian Ocean island of Reunion, were subsequently discovered infested by three species of hyperparsites. These could severely affect the ability of this predator to function as an effective biocontrol agent. Augmentative biological control is a process designed to increase the efficiency of natural enemies already in place. Mass release of *Trichogramma dendrolimi*, an indigenous egg parasitoid of the defoliator *Dendrolimus punctatus* in pine plantations in China is an example of this tactic . Another biological control tactic is the use of biological insecticides such as the bacterial agent, *Bacillus thuringiensis* (Bt), nuclear polyhedrosis viruses or fungal preparations to control insects. Bt is widely used for control of lepidopterous defoliators in both natural and plantation forests. In China and Vietnam, a fungus, *Beauvaria bassiana*, is used for control of the pine caterpillar, *Dendolimus punctatus*.

Mechanical tactics include removal and destruction or rapid removal of infested or infected trees with the objective of destroying the pest. Examples include cutting and burning of trees infested by bark beetles or rapid salvage of infested trees and destruction of infested bark at the sawmill. When the pine woolly aphid, *Pineus borneri*, was first discovered in Kenya, the initial response was to destroy the infested pines. This insect is easily carried on air currents, however, and it was soon realised that the infestation had spread far beyond the designated area where trees were being destroyed.

Use of chemical pesticides applied either from the ground or by low flying aircraft is still considered to be an integral part of IPM. However these materials are used more sparingly, are applied at reduced intervals and with greater precision. In IPM, chemical pesticides are often considered to be the tactic of last resort.

Integration of new technologies

An important part of IPM is that no matter how advanced, sophisticated or effective an IPM system for a specific pest or pest complex may be, there is always room for the introduction of new technologies. These may include more accurate pest monitoring, prediction of new pest management tactics, and more efficacious treatments with fewer undesirable side effects.

Integrated pest management systems

IPM systems consist of a combination of decision-making and pest management tools directed against a pest or pest complex and are in various stages of development. An example of an evolving IPM system is the approach being taken to manage the European wood wasp, *Sirex noctilio,* in pine plantations in southern Brazil.

Early detection of this insect is accomplished by baiting suppressed trees in plantations with an herbicide, a procedure that attracts attacking wasps. Infestations are treated either through inoculation of a parasitic nematode, *Daladenus siricidola,* which renders the female wood wasps incapable of producing eggs, or thinning plantations to reduce stocking and maintain tree vigor.

Damage assessment technologies such as aerial sketch-mapping and aerial photos are currently being evaluated. Landsat satellite imagery is being tested to determine its capacity to map the location of small, non-industrial private pine plantations that might serve as reservoirs for populations of this damaging insect. Introduction of additional parasitic insects is another potential tactic for managing this insect.

REFERENCES

Ciesla, W.M. 1982. IPM: New approaches to old problems. *American Forests* 88(2): 40-44, 51-52.

Ciesla, W.M. 1994. *Assessment of the 1994 outbreak of nun moth, Lymantria monacha in Poland and proposed pest management activities.* FAO, Rome, Italy FO 2/354, 24 pp.

Ciesla, W.M., Mbugua, D.K. & Ward, J.D. 1995. Ensuring forest health and productivity - A perspective from Kenya. *Journal of Forestry* 93(10): 36- 39.

Manion, P.D. 1991. *Tree disease concepts.* Prentice Hall, Engelwood Cliffs, NJ (USA), 402 pp.

Newhook, F.J. & Podger, F.D. 1972. The role of *Phytophthora cinnamomi* in Australian and New Zealand Forests. *Annual Review of Phytopathology* 10: 299-320.

Obiri, J.F. 1994. Variation of cypress aphid, *Cinara cupressi* (Buckton), attack on the family Cupressaceae. *Commonwealth Forestry Review* 73: 43-46.

10

Management of Garden Symphylans

Garden symphylans (*Scutigerella immaculata* Newport) are small, white, "centipede-like" soil arthropods, common in many agricultural production systems. They feed on roots and other subterranean plant parts, causing significant crop losses in some cases. Control can be extremely difficult due to symphylans' vertical movement in the soil, the comlexity of sampling, and the lack of simple, effective control methods. With the recent spread of organic agriculture and better soil management techniques, crop damage associated with symphylans has become more commonplace. It is ironic that these pests are such a problem on farms that practice good soil management – maintaining soil with good tilth, high organic matter, and low compaction.

Diagnosing a garden symphylan problem is sometimes difficult, since damage may be exhibited in a number of forms and garden symphylans are not always easy to find, even when their damage is obvious. Economic damage may result from direct feeding on root and tuber crops and reduced stands of direct-seeded or transplanted crops. However, most commonly, root feeding reduces the crop's ability to take up water and nutrients, which leads to general stunting. Root damage may also render plants more susceptible to some soil-borne plant pathogens. Correct diagnosis of garden symphylan problems and identification of appropriate management tactics for a given cropping system will generally require the following:

1. Sampling to determine whether garden symphylans are present in damaging numbers
2. A general knowledge of management tactics and garden symphylan ecology to select the specific tactics that will be most effective in a given cropping system

IDENTIFICATION OF PESTS

Garden symphylans are not insects, but members of the class Symphyla. Species of this class are common soil arthropods worldwide. Symphyla are small, whitish "centipede-like" creatures ranging from less than 1/8 inch up to about 5/8 inch (or 1/4 inch for garden symphylans). They have 6 to 12 pairs of legs (depending on age), which makes them easy to distinguish from common soil insects (e.g., springtails) and diplurans that have only three pairs of legs, all on the thorax, or middle body segment. Though their color may vary, depending on what they have eaten, garden symphylans are generally whiter and smaller than true centipedes, which are also soil arthropods with many pairs of legs (one pair per body segment) and quick movements. Millipedes are generally slower moving, with two pairs of legs on each body segment.

Figure 1. Garden Symphylan

Symphylans resemble centipedes, but are smaller and translucent. They can move rapidly through the pores between soil particles, and are typically found from the surface down to a depth of about 50 cm. They consume decaying vegetation, but can do considerable harm in an agricultural setting by consuming seeds, roots, and root hairs in cultivated soil.Juveniles have six pairs of legs, but, over a lifetime of several years, add an additional pair at each moult so that the adult instar has twelve pairs of legs. Lacking eyes, their long antennae serve as sense organs. They have several features linking them to early insects, such as a labium (fused second maxillae), an identical number of head segments and certain features of their legs.Some Symphyla species feed primarily on dead or decaying organic matter, playing an important role in cycling nutrients. Other species, such as the garden symphylan, are serious pests, primarily feeding on living plants.

Several Symphyla species are present in the western United States; however, the garden symphylan is the only Symphyla species that is documented to cause crop damage in the western U.S. Garden symphylans are by far the most common Symphyla species found in agricultural systems. If Symphyla are found in an agricultural system at a high density, concentrated around the roots of plants, they are likely to be garden symphylans. The pest is not known to vector any plant diseases, although extensive research on the question has not been conducted.

GARDEN SYMPHYLAN BIOLOGY AND ECOLOGY

In the western U.S., eggs, adults, and immature garden symphylans can be found together throughout most of the year. Temperature plays a key role in regulating oviposition (egg laying), and the greatest numbers of eggs are usually deposited in the spring and fall.

Eggs are pearly white and spherical with hexagonal shaped ridges. Eggs incubate for about 25 to 40 days, when temperatures range from 50° to 70°F, but hatching occurs in about 12 days as temperatures reach 77°F.

First instars emerge from the egg with six pairs of legs and six antennal segments, their bodies covered with fine hairs. Slow movements and a swollen posterior make first instars appear superficially more like a collembolan than an adult garden symphylan. These first instars, however, are rarely found in the rooting zone and within days molt to second instars that resemble small adult garden symphylans.

Each of the six subsequent molts results in the addition of a pair of legs and variable numbers of body and antennal segments. Total time from egg to sexually mature adult (seventh instar) is about five months at 50°F, decreasing to about three months at 70°F and less than two months at 77°F. Therefore, it may be possible to have two complete generations a year. Interestingly, unlike adult insects, which do not molt, adult garden symphylans may molt more than 40 times.

Understanding of garden symphylan occurrence and movement is far from complete. Nonetheless, some generalisations can be made both about soils in which garden symphylans occur more commonly and about their movement in soils.

Garden symphylan populations are highly aggregated within fields and on a larger scale. In Oregon, Washington, and California, garden symphylans are more common in the western regions of the states. Within these regions, garden symphylans tend to occur in heavier irrigated soils, and within these heavier soils, garden symphylans tend to occur in "hotspots" of a few square feet to several acres. Even within shovelfuls of soil, garden symphylans often occur in very distinct aggregations. Garden symphylans are unable to burrow through the soil. They use pores, seasonal cracks and burrows made by other soil animals, such as earthworms, to travel through the soil profile.

In general, practices that improve soil structure (e.g., addition of organic matter, reduced tillage, raised beds) improve the ability of garden symphylans to move through the soil, leading to increased populations and/or increased damage through improved access to roots. As a result, high populations of garden symphylans are more commonly found in fine-textured, heavier soils with moderate or better structure and many macropores, rather than in sandy soils. When garden symphylans are found in sandier soils, these soils have commonly been amended with organic matter.

In the Pacific Northwest and Northern California, garden symphylans are commonly found in alluvial soils, and are likely spread to some extent by flooding. Relative soil acidity does not appear to be closely correlated with garden symphylan occurrence, since symphylans are found in very acid soils (e.g., where blueberries grow) to fairly alkaline soils (e.g., pH 8+).

Hot spots within infested fields often remain consistent from year to year with little change in populations and only minor lateral spread, possibly due to physical characteristics of the soil. However, changes in hotspots do occur.

SYMPHYLAN MOVEMENT IN THE SOIL

If the soil environment is favorable, garden symphylans may migrate from the soil surface to a depth of more than 3 feet. The soil profile, including compacted or sandy horizons and high water tables that may impede movement, determines the depth to which garden symphylans may migrate. Timing of vertical migrations is primarily due to the interaction among moisture, temperature, and internally regulated feeding cycles.

A general understanding of these interactions is important both for the timing and interpretation of sampling efforts and for selecting management tactics. Garden symphylans tend to aggregate in the top six inches of soil when the soil is moist and warm, and move to deeper soil strata when soil becomes very dry or cool.

In Oregon, Washington, and California, garden symphylans are generally found in the surface soil from March through November, with the highest surface populations observed in May and June. Garden symphylans may be found in the surface soil when conditions are fairly warm (e.g., when air temperatures exceed 95°F), if sufficient moisture is present and roots are shallow or absent. In the hottest interior valleys, symphylans may be more active in the spring/early summer and the fall, with surface activity dropping off in July, August, and into September. Garden symphylans migrate to the surface soil to feed, then return to the deeper strata to molt, as demonstrated by the large number of molted skins that are observed in these strata. When garden symphylans are feeding ravenously after molting, they may enter the surface soil zone even in generally unfavorable conditions.

Since migrations are not synchronised, portions of the population are usually present throughout the habitable portion of the soil profile. Presence of garden symphylans in the surface soil may also be influenced by other variables that impede movement, such as tillage and compaction from tractor tires.

SAMPLING

Sampling for garden symphylans is extremely important for identifying damage, for making informed management decisions, and for evaluating the effects of those decisions. Sampling, however, is often difficult. Three main sampling methods are used: baiting, soil sampling, and indirect sampling. Each method has benefits and drawbacks, and the selection of a sampling method will vary, depending on the objectives of the sampling, the time of year, and the site conditions. Part of the difficulty in sampling is due to the patchy distribution of populations. It is important to be aware that an individual sample count provides information only about the region near where that sample was taken.

Counts will often vary from 0 to more than 50 garden symphylans per sample. To get information about the spatial patterns of the population, it is best to take sample units in a grid pattern. Sorting and comparing the samples by site factors such as soil type, drainage, and cropping history may provide valuable information about the distribution of populations within a site. In most cases, sampling measures only the density of symphylans in the surface soil; therefore, sampling should only be conducted when garden symphylans are in the surface soil.

The best sampling conditions are generally warm, moist soil. Sampling within three weeks after major tillage – such as discing, plowing, or spading – is often inaccurate, since garden symphylans may not have had ample time to re-establish in the surface soil. If sampling is conducted soon after tillage, soil sampling methods should be used. Sampling should be conducted to a depth that includes several inches of soil undisturbed by tillage.

Soil Sampling

Soil sampling is the standard/historic method for estimating how many garden symphylans are present in a field (i.e., approximate number of garden symphylans per unit of soil, or estimated population density). Sample unit sizes vary. The most common soil sample units have been the following.

— A 1-foot cube

— A 6-inch square, 1-foot deep

— A "shovelful"

— Cores 3 to 4 inches in diameter and 4 to 12 inches deep.

When soil samples are taken, the soil from each sample unit is usually placed on a piece of dark plastic or cloth, where the aggregates are broken apart and the garden symphylans are counted.

Sampling must be conducted throughout the entire habitable region of the soil profile (i.e., possibly to a depth of more than three feet) to obtain accurate population density estimates, but this is rarely done, because of the extensive time and resources required. Therefore, sampling is usually conducted when garden symphylans are believed to be in the top 6 to 12 inches of the root zone. Shallow sampling (e.g., to a depth of 4 inches) saves time and allows larger areas to be sampled, but deeper sampling (e.g., to a depth of 12 inches) is generally more reliable. Sampling is not recommended in very dry conditions.

Bait Sampling

In recent years, bait sampling methods have been developed. Bait samples are generally much faster to take than soil samples, but they are also more variable and more sensitive to factors such as soil moisture, temperature, and the presence of vegetation. To bait sample, place half of a potato or beet on the soil surface and shelter it with a protective cover. One to three days after placement, lift the bait and count first the garden symphylans on the soil and second the garden symphylans on the bait. During warm and/or dry conditions, baits are generally checked one to two days after placement, as counts decrease if baits are left out for multiple days. In cooler conditions, baits are commonly left out for three to five days. Bait sampling works very well for some applications, though it cannot be used in all conditions.

Baiting works best at least two to three weeks after tillage, when the soil has stabilised, but before plants are well established. If travel to the sampling location requires significant resources, soil sampling methods may be preferred, because they require only one trip to the site.

Indirect Sampling Methods

Plant growth can sometimes be a useful indirect measure of garden symphylan starting point for assessing their spatial patterns. For example, healthy plants sometimes indicate low garden symphylan populations and conversely, unhealthy plants sometimes indicate high garden symphylan populations. Indirect measures such as this may provide valuable information about the extent and pattern of infestation, but they should not be used in place of direct sampling. This is because many factors could lead to healthy plant growth, even within infested soil.

Action Thresholds: Interpretation of Sampling Results

Management decisions, such as those regarding pesticide applications and the intensity of tillage, are sometimes based on pre-planting density estimates of garden symphylan population. Owing partially to the many crops in which garden symphylans are pests, thresholds for individual crops are not well developed. The relationship between garden symphylan population densities and measures such as stand count and yield are influenced by factors such as crop type, tillage intensity, and crop stage.

In the laboratory, levels as low as 5 to 15 garden symphylans per pot have been shown to reduce the growth of seedlings of crops such as snap beans, spinach, and sweet corn. The higher density of 45 garden symphylans per pot has been shown to reduce the growth of tomatoes and spinach seedlings by more than 90%. In the field, noticeable damage often occurs if garden symphylans exceed an average of 5 to 10 per shovelful in moderately to highly susceptible crops such as broccoli, squash, spinach, and cabbage. In conventional cropping systems, two to three garden symphylans per square foot is commonly used as a treatment threshold.

Because of the considerable variability of symphylan densities within a field, sample unit counts may range from 0 to more than 100. These results are helpful in locating field hot spots. In more tolerant crops, such as potatoes, beans, and small grains, garden symphylan feeding may not lead to significant damage, even at considerably higher population densities.

MANAGEMENT AND CONTROL

Many factors influence garden symphylan population levels. However, because it is difficult to accurately sample populations, information about the true effects of many factors on garden symphylans is scant at best. For management purposes it is important to make a distinction between tactics that decrease populations and tactics that reduce damage to crops but may not necessarily decrease populations.

In most cases, effective garden symphylan management involves establishing a balance between these two strategies. It is important to keep in mind that in most cases, after damage is noticed, little can be done without replanting. Sampling is, therefore, important in determining the proper course of action. It is unknown whether symphylan populations may develop from transported soil or compost. These are certainly possible sources of infestation, and it is recommended to sample soil and compost (from on- or off-site) for symphylans before applying these amendments to a field. Generally, symphylan populations are thought to be home-grown and to develop over time due to favorable soil management practices.

Tactics to Decrease Populations

Reducing populations has been the focus of many studies. Though no "silver bullets" have been identified, some tactics are available. Probably no method will eradicate garden symphylans from a site, and the effect of most tactics will not last longer than one to three years.

Tillage

Tillage is probably the oldest control tactic, and it is still one of the most effective. Tillage can physically crush garden symphylans, thus reducing populations. Tillage may also harm populations of key garden symphylan predators such as centipedes and predaceous mites. However, in annual cropping systems, the benefits of increased predator populations from reduced tillage have not been shown to be as effective as tillage in decreasing garden symphylan populations. When considering increased tillage, farmers need to balance the benefits and the costs, such as oxidised soil organic matter, compacted soil, and increased expenses for time, fuel, and worn equipment. In general, for the most effective control, till when the garden symphylans are in the surface soil and when soil moisture allows preparation of a fine seed bed. Since only a portion of the symphylan population is in the surface soil, tillage never provides complete control. However, surface populations are generally significantly lower for at least two to three weeks after tillage.

Pesticides

Hundreds of compounds have been used against garden symphylans in the past 100 years, with varying efficacies. Fumigants and organophosphate pesticides have been the most effective, but many of these are no longer registered. Pesticides may have the effect of both killing garden symphylans and repelling them from the surface soil. Less toxic pesticides (e.g., pyrethroids and other natural pesticides) have not been shown to provide acceptable control. Pesticides generally provide the greatest amount of control when they are broadcast and incorporated, though banded and injected applications can provide an acceptable level of control.

Crop rotation

Although garden symphylans feed on a wide range of plants, they can also persist in bare soil by feeding on other soil organisms. Plants vary greatly in their susceptibility to garden symphylans. Crop rotation may partially explain seemingly sudden shifts in garden symphylan population levels. Populations decrease significantly in potato crops, even allowing subsequent cultivation of susceptible crops in rotation. Though no other crops have been shown to be nearly as effective at reducing symphylan populations as potatoes, symphylan populations are lower after a spring oat ('Monida') winter cover crop than after a mustard ('Martiginia'), barley ('Micah'), or rye ('Wheeler') winter cover crop. Mustard and spinach crops are very good hosts and may lead to increased populations in some cases. All these factors should be considered when developing a weed management plan.

Other Soil Amendments

The reported effects of common soil amendments such as manure, lime, fertilisers, and compost vary greatly and are often contradictory. Lime and fertilisers are generally accepted to have little effect on populations, while manure applications are generally believed to increase populations. The effect of compost and organic amendments on garden symphylan populations has been variable, but at this point none have been shown to consistently and significantly reduce garden symphylan populations.

Tactics to Reduce Damage to Crops

Most plants can tolerate some level of garden symphylan feeding during all or part of the growing season, and two general tactics can help to grow healthy crops in symphylan-infested soil. These tactics are those aimed at reducing crop damage when garden symphylan populations are high, and those aimed at reducing the number of garden symphylans on crop roots during establishment, when plants are often most susceptible.

Tactics to Reduce Crop Damage

Crop species/variety

Susceptibility to garden symphylan feeding can vary dramatically between different plant species and varieties. In most cases tolerance to feeding seems due to increased vigor and/or root production (e.g., broccoli, corn). In some cases garden symphylans may simply eat less of certain crops/varieties, though this has not been demonstrated experimentally. Generally, smaller-seeded crops tend to be more susceptible than larger-seeded crops. Commonly damaged crops include broccoli and other cole crops, spinach, beets, onions, and squash. Beans and potatoes are rarely damaged, even under high populations. Perennial crops, such as strawberries, raspberries, hops, and bare root trees (nursery production), can also be damaged, particularly during establishment. Within a crop species, such as broccoli, some varieties are more tolerant of garden symphylans than others.

Crop stage

Within a crop, susceptibility is often related to the developmental stage of the crop. For example, within a tomato variety, direct-seeded tomatoes are more susceptible than 4-week-old transplants, which are more susceptible than 12-week-old transplants. Using transplants or increasing transplant size to reduce damage is not effective for all crops. Transplants of broccoli and eggplant, for example, often fail to establish under high garden symphylan populations.

Plant density

Garden symphylans do not cross the soil surface for significant distances, as do ground beetles. However, they are quite active and surprisingly mobile for their size, moving vertically and horizontally through the soil profile. This is strikingly evident when, for example, seedlings transplanted into a stale seedbed with seemingly few garden symphylans have garden symphylans crawling all over their roots less than one day after planting.

The number of garden symphylans feeding on each plant in a local region (e.g., raised bed) is partially a factor of the number of plants present in that bed. In some cases, increasing plant density – which of course must be balanced with plant competition considerations – brings about improved production. Modifications of this strategy include planting an early "distraction" or "dilution" crop in a bed or adjacent to a cropping row. A good dilution crop is a low-cost, vigorous, easy-establishing crop (e.g., sudangrass in suitable conditions) that increases the roots in the soil and effectively "dilutes" the garden symphylans enough to get the target crop established. The dilution crop is then removed as the target crop establishes.

Tactics to Reduce Access of Garden Symphylans to Crop Roots

Since garden symphylans are not able to burrow through soil, instead relying on soil pores and channels made by roots and other soil organisms, their access to roots is strongly correlated with soil structure, bulk density ("fluffiness") of the soil, and pore connectivity. In general, the following tactics focus on temporarily reducing the number of garden symphylans in the surface soil before planting, thus allowing crops to establish while garden symphylan numbers are low.

Tillage

Along with directly killing garden symphylans, tillage breaks apart soil aggregates, modifying soil pores and pore connectivity. The effects of tillage vary with the types of implements used. In general, the more disruptive the tillage, the greater the effect it will have on garden symphylans. Plowing or discing, followed by thorough preparation of a fine seed bed using a rototiller or roterra, often reduces surface-feeding garden symphylan populations for two to three weeks. Over this period of time, pores are formed as some aggregation occurs, and earthworms and plant roots make new channels through the soil. Less intense soil disturbance, such as hand digging or shallow cultivation with a harrow or strip tiller, may have a significantly less disruptive effect on garden symphylans.

Compaction/Raised Beds

The protection of plant roots from garden symphylans is sometimes evident in zones where tractor tires have compacted the soil, or in areas where a rototiller or disc has formed a compacted layer or "plow pan." Although compaction can have some negative effects, in some soils it is possible to compact the soil beneficially using, for example, a landscaping roller, thus reducing garden symphylan movement enough to allow plants to establish. The opposite conditions often occur in raised beds that are highly amended with organic matter, where the soil is very low in bulk density and garden symphylans are able to move freely throughout the beds.

REFERENCES

Berry, R. E. 1972. Garden symphylan: Reproduction and development in the laboratory. (*Scutigerella immaculata*). Journal of Economic Entomology. Vol. 65. p. 1628-1632.

Edwards, C. A. 1957. Simple techniques for rearing Collembola, Symphyla and other small soil-inhabiting arthropods, pp. 412-416. In D. K. M. Kevan [ed.], Soil Zoology. Butterworths Publications Ltd., London.

Filinger, G. A. 1928. Observations on the habits and control of the garden centipede, *Scutigerella immaculata,* Newport, a pest in greenhouses. J. Econ. Entomol. 2: 357-360.

Filinger, G. A. 1931. The Garden Symphylid. Ohio Agr. Exp. Sta. Bul. 486: 1-33.

Getzin, L. W., and C. H. Shanks. 1964. Infection of the garden symphylan, *Scutigerella immaculata* (Newport), by *Entomophthora coronata* (Constantin) Kevorkian and *Metarrhizium anisopliae* (Metchnikoff) Sorokin. J. Insect Path. 6: 542-543.

Gould, G. E., and C. A. Edwards. 1968. Damage to field corn by symphylans. Proc. Indiana Acad. Sci. 77: 214-221.

Howitt, A. J. 1959. Control of *Scutigerella immaculata* (Newport) in the Pacific Northwest with soil fumigants. J. Econ. Entomol. 52: 678-683.

Illingworth, J. F. 1927. Symphylids destructive to the roots of pineapple. Pineapple News Mar-Dec: 88-91.

Koontz, F. R. 1968. Biological and ecological relationships of the fungus, *Entomophthora coronata* (Constantin) Kevorkian, and the garden symphylan, *Scutigerella immaculata* (Newport). Ph.D. dissertation, Oregon State University, Corvallis.

11

Biointensive Pest Management

Pest management is an ecological matter. The size of a pest population and the damage it inflicts is, to a great extent, a reflection of the design and management of a particular agricultural ecosystem. We humans compete with other organisms for food and fiber from our crops. We wish to secure a maximum amount of the food resource from a given area with minimum input of resources and energy.

However, if the agricultural system design and/or management is faulty—making it easy for pests to develop and expand their populations or, conversely, making it difficult for predators and parasites of pests to exist—then we will be expending unnecessary resources for pest management.

Therefore, the first step in sustainable and effective pest management is looking at the design of the agricultural ecosystem and considering what ecological concepts can be applied to the design and management of the system to better manage pests and their parasites and predators. The design and management of our agricultural systems need re-examining. We've come to accept routine use of biological poisons in our food systems as normal.

But routine use of synthetic chemicals represents significant energy inputs into the agricultural system, and carries both obvious and hidden costs to the farmer and society. Attempting to implement an ecology-based discipline like Integrated Pest Management (IPM) in large monocultures, which substitute chemical inputs for ecological design, can be an exercise in futility and inefficiency.

However, IPM has strayed from its ecological roots. Critics of what might be termed "conventional" IPM note that it has been implemented as Integrated Pesticide Management with an emphasis on using pesticides as a tool of first resort. What has been missing from this approach, which is essentially reactive, is an understanding of the ecological basis of pest infestations.

AIMS OF BIOINTENSIVE IPM

Biointensive IPM incorporates ecological and economic factors into agricultural system design and decision making, and addresses public concerns about environmental quality and food safety. The benefits of implementing biointensive IPM can include reduced chemical input costs, reduced on-farm and off-farm environmental impacts, and more effective and sustainable pest management.

An ecology-based IPM has the potential of decreasing inputs of fuel, machinery, and synthetic chemicals—all of which are energy intensive and increasingly costly in terms of financial and environmental impact. Such reductions will benefit the grower and society. Over-reliance on the use of synthetic pesticides in crop protection programs around the world has resulted in disturbances to the environment, pest resurgence, pest resistance to pesticides, and lethal and sub-lethal effects on non-target organisms, including humans.

These side effects have raised public concern about the routine use and safety of pesticides. At the same time, population increases are placing ever-greater demands upon the "ecological services"—that is, provision of clean air, water and wildlife habitat—of a landscape dominated by farms. Although some pending legislation has recognised the costs to farmers of providing these ecological services, it's clear that farmers and ranchers will be required to manage their land with greater attention to direct and indirect off-farm impacts of various farming practices on water, soil, and wildlife resources. With this likely future in mind, reducing dependence on chemical pesticides in favor of ecosystem manipulations is a good strategy for farmers.

Consumers Union, a group that has carried out research and advocacy on various pesticide problems for many years, defines biointensive IPM as the highest level of IPM:

— A systems approach to pest management based on an understanding of pest ecology. It begins with steps to accurately diagnose the nature and source of pest problems, and then relies on a range of preventive tactics and biological controls to keep pest populations within acceptable limits. Reduced-risk pesticides are used if other tactics have not been adequately effective, as a last resort, and with care to minimise risks.

This "biointensive" approach sounds remarkably like the original concept of IPM. Such a "systems" approach makes sense both intuitively and in practice.

The primary goal of biointensive IPM is to provide guidelines and options for the effective management of pests and beneficial organisms in an ecological context. The flexibility and environmental compatibility of a biointensive IPM strategy make it useful in all types of cropping systems.

Even conventional IPM strategies help to prevent pest problems from developing, and reduce or eliminate the use of chemicals in managing problems that do arise. Results of 18 economic evaluations of conventional IPM on cotton showed a decrease in production costs of 7 percent and an average decrease in pesticide use of 15 percent. Biointensive IPM would likely decrease chemical use and costs even further. Prior to the mid-1970s, lygus bugs were considered to be the key pest in California cotton. Yet in large-scale studies on insecticidal control of lygus bugs, yields in untreated plots were not significantly different from those on treated plots. This was because the insecticides often induced outbreaks of secondary lepidopterous larvae and mite pests which caused additional damage as well as pest resurgence of the lygus bug itself.

These results, from an economic point of view, seem paradoxical, as the lygus bug treatments were costly, yet the treated plots consistently had lower yields. This paradox was first pointed out by R. van den Bosch, V. Stern, and L. A. Falcon, who forced a reevaluation of the economic basis of Lygus control in California cotton.

COMPONENTS OF BIOINTENSIVE IPM

An important difference between conventional and biointensive IPM is that the emphasis of the latter is on proactive measures to redesign the agricultural ecosystem to the disadvantage of a pest and to the advantage of its parasite and predator complex. At the same time, biointensive IPM shares many of the same components as conventional IPM, including monitoring, use of economic thresholds, record keeping, and planning.

Good planning must precede implementation of any IPM program, but is particularly important in a biointensive program. Planning should be done before planting because many pest strategies require steps or inputs, such as beneficial organism habitat management, that must be considered well in advance. Attempting to jump-start an IPM program in the beginning or middle of a cropping season generally does not work.

When planning a biointensive IPM program, some considerations include:

- Options for design changes in the agricultural system (beneficial organism habitat, crop rotations).
- Choice of pest-resistant cultivars.
- Technical information needs.
- Monitoring options, record keeping, equipment, etc.

The Pest Manager/Ecosystem Manager

The pest manager is the most important link in a successful IPM program. The manager must know the biology of the pest and the beneficial organisms associated with the pest, and understand their interactions within the farm environment. As a detailed knowledge of the pest is developed, weak links in its life cycle become apparent. These weak links are phases of the life cycle when the pest is most susceptible to control measures. The manager must integrate this knowledge with tools and techniques of biointensive IPM to manage not one, but several pests.

A more accurate title for the pest manager is "ecosystem doctor," for he or she must pay close attention to the pulse of the managed ecosystem and stay abreast of developments in IPM and crop/pest biology and ecology. In this way, the ecosystem manager can take a proactive approach to managing pests, developing ideas about system manipulations, testing them, and observing the results. IPM options may be considered proactive or reactive. Proactive options, such as crop rotations and creation of habitat for beneficial organisms, permanently lower the carrying capacity of the farm for the pest.

The carrying capacity is determined by factors like food, shelter, natural enemies complex, and weather, which affect the reproduction and survival of a species. Cultural controls are generally considered to be proactive strategies. The second set of options is more reactive. This simply means that the grower responds to a situation, such as an economically damaging population of pests, with some type of short-term suppressive action. Reactive methods generally include inundative releases of biological controls, mechanical and physical controls, and chemical controls.

Proactive Strategies

- Healthy, biologically active soils (increasing belowground diversity)
- Habitat for beneficial organisms (increasing aboveground diversity)
- Appropriate plant cultivars

Cultural controls are manipulations of the agroecosystem that make the cropping system less friendly to the establishment and proliferation of pest populations. Although they are designed to have positive effects on farm ecology and pest management, negative impacts may also result, due to variations in weather or changes in crop management.

Maintaining and increasing biological diversity of the farm system is a primary strategy of cultural control. Decreased biodiversity tends to result in agroecosystems that are unstable and prone to recurrent pest outbreaks and many other problems. Systems high in biodiversity tend to be more "dynamically stable"—that is, the variety of organisms provide more checks and balances on each other, which helps prevent one species (i.e., pest species) from overwhelming the system.

There are many ways to manage and increase biodiversity on a farm, both above ground and in the soil. In fact, diversity above ground influences diversity below ground. Research has shown that up to half of a plant's photosynthetic production (carbohydrates) is sent to the roots, and half of that (along with various amino acids and other plant products) leaks out the roots into the surrounding soil, providing a food source for microorganisms. These root exudates vary from plant species to plant species and this variation influences the type of organisms associated with the root exudates.

Factors influencing the health and biodiversity of soils include the amount of soil organic matter; soil pH; nutrient balance; moisture; and parent material of the soil. Healthy soils with a diverse community of organisms support plant health and nutrition better than soils deficient in organic matter and low in species diversity. Research has shown that excess nutrients as well as relative nutrient balance (i.e., ratios of nutrients—for example, twice as much calcium as magnesium, compared to equal amounts of both) in soils affect insect pest response to plants.

Imbalances in the soil can make a plant more attractive to insect pests, less able to recover from pest damage, or more susceptible to secondary infections by plant pathogens. Soils rich in organic matter tend to suppress plant pathogens. In addition, it is estimated that 75% of all insect pests spend part of their life cycle in the soil, and many of their natural enemies occur there as well. For example, larvae of one species of blister beetle consume about 43 grasshopper eggs before maturing. Both are found in the soil. Overall, a healthy soil with a diversity of beneficial organisms and high organic matter content helps maintain pest populations below their economic thresholds.

Genetic diversity of a particular crop may be increased by planting more than one cultivar. For example, a recent experiment in China demonstrated that disease-susceptible rice varieties planted in mixtures with resistant varieties had 89% greater yield and a 94% lower incidence of rice blast (a fungus) compared to when they were grown in monoculture. The experiment, which involved five townships in 1998 and ten townships in 1999, was so successful that fungicidal sprays were no longer applied by the end of the two-year program.

Species diversity of the associated plant and animal community can be increased by allowing trees and other native plants to grow in fence rows or along water ways, and by integrating livestock into the farm system. Use of the following cropping schemes offers additional ways to increase species diversity.

Crop rotations radically alter the environment both above and below ground, usually to the disadvantage of pests of the previous crop. The same crop grown year after year on the same field will inevitably build up populations of organisms that feed on that plant, or, in the case of weeds, have a life cycle similar to that of the crop. Add to this the disruptive effect of pesticides on species diversity, both above and below ground, and the result is an unstable system in which slight stresses (e.g., new pest variety or drought) can devastate the crop.

An enforced rotation program in the Imperial Valley of California has effectively controlled the sugar beet cyst nematode. Under this program, sugar beets may not be grown more than two years in a row or more than four years out of ten in clean fields (i.e., non-infested fields). In infested fields, every year of a sugar beet crop must be followed by three years of a non-host crop. Other nematode pests commonly controlled with crop rotation methods include the golden nematode of potato, many root-knot nematodes, and the soybean cyst nematode.

When making a decision about crop rotation, consider the following questions: Is there an economically sustainable crop that can be rotated into the cropping system? Is it compatible? Important considerations when developing a crop rotation are:

- What two crops can provide an economic return when considered together as a biological and economic system that includes considerations of sustainable soil management?
- What are the impacts of this season's cropping practices on subsequent crops?
- What specialised equipment is necessary for the crops?
- What markets are available for the rotation crops?

A corn/soybean rotation is one example of rotating compatible economic crops. Corn is a grass; soybean is a leguminous broadleaf. The pest complex of each, including soil organisms, is quite different. Corn rootworm, one of the major pests of corn, is virtually eliminated by using this rotation. Both crops generally provide a reasonable return. Even rotations, however, create selection pressures that will ultimately alter pest genetics. A good example is again the corn rootworm: the corn/bean rotation has apparently selected for a small population that can survive a year of non-corn (i.e., soybean) cropping.

Management factors should also be considered. For example, one crop may provide a lower direct return per acre than the alternate crop, but may also lower management costs for the alternate crop, with a net increase in profit.

Other Cropping Structure Options

Multiple cropping is the sequential production of more than one crop on the same land in one year. Depending on the type of cropping sequence used, multiple cropping can be useful as a weed control measure, particularly when the second crop is interplanted into the first.

Interplanting is seeding or planting a crop into a growing stand, for example overseeding a cover crop into a grain stand.

There may be microclimate advantages (e.g., timing, wind protection, and less radical temperature and humidity changes) as well as disadvantages (competition for light, water, nutrients) to this strategy. By keeping the soil covered, interplanting may also help protect soil against erosion from wind and rain. Intercropping is the practice of growing two or more crops in the same, alternate, or paired rows in the same area. This technique is particularly appropriate in vegetable production.

The advantage of intercropping is that the increased diversity helps "disguise" crops from insect pests, and if done well, may allow for more efficient utilisation of limited soil and water resources. Disadvantages may relate to ease of managing two different crop species— with potentially different nutrient, water, and light needs, and differences in harvesting time and method—in close proximity to each other. Strip cropping is the practice of growing two or more crops in different strips across a field wide enough for independent cultivation.

It is commonly practiced to help reduce soil erosion in hilly areas. Like intercropping, strip cropping increases the diversity of a cropping area, which in turn may help "disguise" the crops from pests. Another advantage to this system is that one of the crops may act as a reservoir and/or food source for beneficial organisms. However, much more research is needed on the complex interactions between various paired crops and their pest/predator complexes. The options described above can be integrated with no-till cultivation schemes and all its variations as well as with hedgerows and intercrops designed for beneficial organism habitat. With all the cropping and tillage options available, it is possible, with creative and informed management, to evolve a biologically diverse, pest-suppressive farming system appropriate to the unique environment of each farm.

Other Cultural Management Options

Disease-free seed and plants are available from most commercial sources, and are certified as such. Use of disease-free seed and nursery stock is important in preventing the introduction of disease.

Resistant varieties are continually being bred by researchers. Growers can also do their own plant breeding simply by collecting non-hybrid seed from healthy plants in the field. The plants from these seeds will have a good chance of being better suited to the local environment and of being more resistant to insects and diseases. Since natural systems are dynamic rather than static, breeding for resistance must be an ongoing process, especially in the case of plant disease, as the pathogens themselves continue to evolve and become resistant to control measures. Sanitation involves removing and destroying the overwintering or breeding sites of the pest as well as preventing a new pest from establishing on the farm. This strategy has been particularly useful in horticultural and tree-fruit crop situations involving twig and branch pests.

If, however, sanitation involves removal of crop residues from the soil surface, the soil is left exposed to erosion by wind and water. As with so many decisions in farming, both the short- and long-term benefits of each action should be considered when tradeoffs like this are involved.

Spacing of plants heavily influences the development of plant diseases and weed problems. The distance between plants and rows, the shape of beds, and the height of plants influence air flow across the crop, which in turn determines how long the leaves remain damp from rain and morning dew. Generally speaking, better air flow will decrease the incidence of plant disease. However, increased air flow through wider spacing will also allow more sunlight to the ground, which may increase weed problems. This is another instance in which detailed knowledge of the crop ecology is necessary to determine the best pest management strategies. How will the crop react to increased spacing between rows and between plants? Will yields drop because of reduced crop density? Can this be offset by reduced pest management costs or fewer losses from disease?

Altered planting dates can at times be used to avoid specific insects, weeds, or diseases. For example, squash bug infestations on cucurbits can be decreased by the delayed planting strategy, i.e., waiting to establish the cucurbit crop until overwintering adult squash bugs have died. To assist with disease management decisions, the Cooperative Extension Service (CES) will often issue warnings of "infection periods" for certain diseases, based upon the weather.

In some cases, the CES also keeps track of "degree days" needed for certain important insect pests to develop. Insects, being cold-blooded, will not develop below or above certain threshold temperatures. Calculating accumulated degree days, that is, the number of days above the threshold development temperature for an insect pest, makes the prediction of certain events, such as egg hatch, possible. University of California has an excellent Web site that uses weather station data from around the state to help California growers predict pest emergence.

Some growers gauge the emergence of insect pests by the flowering of certain non-crop plant species native to the farm. This method uses the "natural degree days" accumulated by plants. For example, a grower might time cabbage planting for three weeks after the *Amelanchier* species on their farm are in bloom. This will enable the grower to avoid peak egg-laying time of the cabbage maggot fly, as the egg hatch occurs about the time *Amelanchier* species are flowering. Using this information, cabbage maggot management efforts could be concentrated during a known time frame when the early instars are active.

Optimum growing conditions are always important. Plants that grow quickly and are healthy can compete with and resist pests better than slow-growing, weak plants. Too often, plants grown outside their natural ecosystem range must rely on pesticides to overcome conditions and pests to which they are not adapted.

Mulches, living or non-living, are useful for suppression of weeds, insect pests, and some plant diseases. Hay and straw, for example, provide habitat for spiders. Research in Tennessee showed a 70% reduction in damage to vegetables by insect pests when hay or straw was used as mulch.

The difference was due to spiders, which find mulch more habitable than bare ground. Other researchers have found that living mulches of various clovers reduce insect pest damage to vegetables and orchard crops. Again, this reduction is due to natural predators and parasites provided habitat by the clovers. Vetch has been used as both a nitrogen source and as a weed suppressive mulch in tomatoes in Maryland. Growers must be aware that mulching may also provide a more friendly environment for slugs and snails, which can be particularly damaging at the seedling stage.

Mulching helps to minimise the spread of soil-borne plant pathogens by preventing their transmission through soil splash. Mulch, if heavy enough, prevents the germination of many annual weed seeds. Winged aphids are repelled by silver- or aluminum-colored mulches. Recent springtime field tests at the Agricultural Research Service in Florence, South Carolina, have indicated that red plastic mulch suppresses root-knot nematode damage in tomatoes by diverting resources away from the roots (and nematodes) and into foliage and fruit.

Biotech Crops. Gene transfer technology is being used by several companies to develop cultivars resistant to insects, diseases, and herbicides. An example is the incorporation of genetic material from *Bacillus thuringiensis* (Bt), a naturally occurring bacterium, into cotton, corn, and potatoes, to make the plant tissues toxic to bollworm, earworm, and potato beetle larvae, respectively.

Whether or not this technology should be adopted is the subject of much debate. Opponents are concerned that by introducing Bt genes into plants, selection pressure for resistance to the Bt toxin will intensify and a valuable biological control tool will be lost. There are also concerns about possible impacts of genetically-modified plant products (i.e., root exudates) on non-target organisms as well as fears of altered genes being transferred to weed relatives of crop plants. Whether there is a market for gene-altered crops is also a consideration for farmers and processors. Proponents of this technology argue that use of such crops decreases the need to use toxic chemical pesticides.

BIOLOGICAL CONTROLS

Biological control is the use of living organisms—parasites, predators, or pathogens—to maintain pest populations below economically damaging levels, and may be either natural or applied. A first step in setting up a biointensive IPM program is to assess the populations of beneficials and their interactions within the local ecosystem. This will help to determine the potential role of natural enemies in the managed agricultural ecosystem. It should be noted that some groups of beneficials may be absent or scarce on some farms because of lack of habitat. These organisms might make significant contributions to pest management if provided with adequate habitat.

Natural biological control results when naturally occurring enemies maintain pests at a lower level than would occur without them, and is generally characteristic of biodiverse systems. Mammals, birds, bats, insects, fungi, bacteria, and viruses all have a role to play as predators and parasites in an agricultural system. By their very nature, pesticides decrease the biodiversity of a system, creating the potential for instability and future problems. Pesticides, whether synthetically or botanically derived, are powerful tools and should be used with caution.

Creation of habitat to enhance the chances for survival and reproduction of beneficial organisms is a concept included in the definition of natural biocontrol. Farmscaping is a term coined to describe such efforts on farms. Habitat enhancement for beneficial insects, for example, focuses on the establishment of flowering annual or perennial plants that provide pollen and nectar needed during certain parts of the insect life cycle.

Other habitat features provided by farmscaping include water, alternative prey, perching sites, overwintering sites, and wind protection. Beneficial insects and other beneficial organisms should be viewed as mini-livestock, with specific habitat and food needs to be included in farm planning. The success of such efforts depends on knowledge of the pests and beneficial organisms within the cropping system. Where do the pests and beneficials overwinter? What plants are hosts and non-hosts? When this kind of knowledge informs planning, the ecological balance can be manipulated in favor of beneficials and against the pests.

It should be kept in mind that ecosystem manipulation is a two-edged sword. Some plant pests are attracted to the same plants that attract beneficials. The development of beneficial habitats with a mix of plants that flower throughout the year can help prevent such pests from migrating *en masse* from farmscaped plants to crop plants.

Applied biological control, also known as augmentative biocontrol, involves supplementation of beneficial organism populations, for example through periodic releases of parasites, predators, or pathogens. This can be effective in many situations—well-timed inundative releases of *Trichogramma* egg wasps for codling moth control, for instance.

Most of the beneficial organisms used in applied biological control today are insect parasites and predators. They control a wide range of pests from caterpillars to mites. Some species of biocontrol organisms, such as *Eretmocerus californicus*, a parasitic wasp, are specific to one host—in this case the sweetpotato whitefly. Others, such as green lacewings, are generalists and will attack many species of aphids and whiteflies.

Information about rates and timing of release is available from suppliers of beneficial organisms. It is important to remember that released insects are mobile; they are likely to leave a site if the habitat is not conducive to their survival. Food, nectar, and pollen sources can be "farmscaped" to provide suitable habitat.

The quality of commercially available applied biocontrols is another important consideration. For example, if the organisms are not properly labeled on the outside packaging, they may be mishandled during transport, resulting in the death of the organisms. A recent study by Rutgers University noted that only two of six suppliers of beneficial nematodes sent the expected numbers of organisms, and only one supplier out of the six provided information on how to assess product viability.

While augmentative biocontrols can be applied with relative ease on small farms and in gardens, applying some types of biocontrols evenly over large farms has been problematic. New mechanised methods that may improve the economics and practicality of large-scale augmentative biocontrol include ground application with "biosprayers" and aerial delivery using small-scale (radio-controlled) or conventional aircraft. Inundative releases of beneficials into greenhouses can be particularly effective. In the controlled environment of a greenhouse, pest infestations can be devastating; there are no natural controls in place to suppress pest populations once an infestation begins.

For this reason, monitoring is very important. If an infestation occurs, it can spread quickly if not detected early and managed. Once introduced, biological control agents cannot escape from a greenhouse and are forced to concentrate predation/parasitism on the pest(s) at hand. An increasing number of commercially available biocontrol products are made up of microorganisms, including fungi, bacteria, nematodes, and viruses.

MECHANICAL AND PHYSICAL CONTROLS

Methods included in this category utilise some physical component of the environment, such as temperature, humidity, or light, to the detriment of the pest. Common examples are tillage, flaming, flooding, soil solarisation, and plastic mulches to kill weeds or to prevent weed seed germination.

Heat or steam sterilisation of soil is commonly used in greenhouse operations for control of soil-borne pests. Floating row covers over vegetable crops exclude flea beetles, cucumber beetles, and adults of the onion, carrot, cabbage, and seed corn root maggots. Insect screens are used in greenhouses to prevent aphids, thrips, mites, and other pests from entering ventilation ducts. Large, multi-row vacuum machines have been used for pest management in strawberries and vegetable crops.

Pest Identification

A crucial step in any IPM program is to identify the pest. The effectiveness of both proactive and reactive pest management measures depend on correct identification. Misidentification of the pest may be worse than useless; it may actually be harmful and cost time and money. After a pest is identified, appropriate and effective management depends on knowing answers to a number — What plants are hosts and non-hosts of this pest?

- When does the pest emerge or first appear?
- Where does it lay its eggs? In the case of weeds, where is the seed source? For plant pathogens, where is the source(s) of inoculum?
- Where, how, and in what form does the pest overwinter?
- How might the cropping system be altered to make life more difficult for the pest and easier for its natural controls?

Monitoring (field scouting) and economic injury and action levels are used to help answer these and additional questions.

Monitoring

Monitoring involves systematically checking crop fields for pests and beneficials, at regular intervals and at critical times, to gather information about the crop, pests, and natural enemies. Sweep nets, sticky traps, and pheromone traps can be used to collect insects for both identification and population-density information. Leaf counts are one method for recording plant growth stages. Square-foot or larger grids laid out in a field can provide a basis for comparative weed counts. Records of rainfall and temperature are sometimes used to predict the likelihood of disease infections.

Specific scouting methods have been developed for many crops. The Cooperative Extension Service can provide a list of IPM manuals available in each state. Many resources are now available via Internet. The more often a crop is monitored, the more information the grower has about what is happening in the fields. Monitoring activity should be balanced against its costs. Frequency may vary with temperature, crop, growth phase of the crop, and pest populations. If a pest population is approaching economically damaging levels, the grower will want to monitor more frequently.

Economic Injury and Action Levels

The economic injury level (EIL) is the pest population that inflicts crop damage greater than the cost of control measures. Because growers will generally want to act before a population reaches EIL, IPM programs use the concept of an economic threshold level (ETL or ET), also known as an action threshold. The ETL is closely related to the EIL, and is the point at which suppression tactics should be applied in order to prevent pest populations from increasing to injurious levels.

In practice, many crops have no established EILs or ETLs, or the EILs that have been developed may be static over the course of a season and thus not reflect the changing nature of the agricultural ecosystem. For example, a single cutworm can do more damage to an emerging cotton plant than to a plant that is six weeks old. Clearly, this pest's EIL will change as the cotton crop develops. ETLs are intimately related to the value of the crop and the part of the crop being attacked. For example, a pest that attacks the fruit or vegetable will have a much lower ETL than a pest that attacks a non-saleable part of the plant.

The exception to this rule is an insect or nematode pest that is also a disease vector. Depending on the severity of the disease, the grower may face a situation where the ETL for a particular pest is zero, i.e., the crop cannot tolerate the presence of a single pest of that particular species because the disease it transmits is so destructive.

SPECIAL CONSIDERATIONS

Cosmetic Damage and Aesthetics

Consumer attitudes toward how produce looks is often a major factor when determining a crop's sale price. Cosmetic damage is an important factor when calculating the EIL, since pest damage, however superficial, lowers a crop's market value. Growers selling to a market that is informed about IPM or about organically grown produce may be able to tolerate higher levels of cosmetic damage to their produce.

Record-keeping

Monitoring goes hand-in-hand with record-keeping, which forms the collective "memory" of the farm. Records should not only provide information about when and where pest problems have occurred, but should also incorporate information about cultural practices (irrigation, cultivation, fertilisation, mowing, etc.) and their effect on pest and beneficial populations. The effects of non-biotic factors, especially weather, on pest and beneficial populations should also be noted. Record-keeping is simply a systematic approach to learning from experience. A variety of software programs are now available to help growers keep track of—and access—data on their farm's inputs and outputs.

Time and Resources

A successful biointensive IPM program takes time, money, patience, short- and long-term planning, flexibility, and commitment. The pest manager must spend time on self-education and on making contacts with Extension and research personnel. Be aware that some IPM strategies, such as increasing beneficial insect habitat, may take more than a year to show results. A well-run biointensive IPM system may require a larger initial outlay in terms of time and money than a conventional IPM program. In the long run, however, a good biointensive IPM program should pay for itself. Direct pesticide application costs are saved and equipment wear and tear may be reduced.

Chemical Controls

Included in this category are both synthetic pesticides and botanical pesticides. Synthetic pesticides comprise a wide range of man-made chemicals used to control insects, mites, weeds, nematodes, plant diseases, and vertebrate and invertebrate pests. These powerful chemicals are fast acting and relatively inexpensive to purchase.

Pesticides are the option of last resort in IPM programs because of their potential negative impacts on the environment, which result from the manufacturing process as well as from their application on the farm. Pesticides should be used only when other measures, such as biological or cultural controls, have failed to keep pest populations from approaching economically damaging levels.

If chemical pesticides must be used, it is to the grower's advantage to choose the least-toxic pesticide that will control the pest but not harm non-target organisms such as birds, fish, and mammals. Pesticides that are short-lived or act on one or a few specific organisms are in this class. Examples include insecticidal soaps, horticultural oils, copper compounds (e.g., bordeaux mix), sulfur, boric acid, and sugar esters.

Biorational pesticides. Although use of this term is relatively common, there is no legally accepted definition. (24) Biorational pesticides are generally considered to be derived from naturally occurring compounds or are formulations of microorganisms. Biorationals have a narrow target range and are environmentally benign. Formulations of Bacillus thuringiensis, commonly known as Bt, are perhaps the best-known biorational pesticide. Other examples include silica aerogels, insect growth regulators, and particle film barriers.

Particle film barriers. A relatively new technology, particle film barriers are currently available under the tradename Surround® WP Crop Protectant. The active ingredient is kaolin clay, an edible mineral long used as an anti-caking agent in processed foods, and in such products as toothpaste and Kaopectate. There appears to be no mammalian toxicity or any danger to the environment posed by the use of kaolin in pest control. The kaolin in Surround is processed to a specific particle size range, and combined with a sticker-spreader. Non-processed kaolin clay may be phytotoxic.

Surround is sprayed on as a liquid, which evaporates, leaving a protective powdery film on the surfaces of leaves, stems, and fruit. Conventional spray equipment can be used and full coverage is important. The film works to deter insects in several ways. Tiny particles of the clay attach to the insects when they contact the plant, agitating and repelling them. Even if particles don't attach to their bodies, the insects may find the coated plant or fruit unsuitable for feeding and egg-laying. In addition, the highly reflective white coating makes the plant less recognisable as a host.

Sugar Esters. Throughout four years of tests, sugar esters have performed as well as or better than conventional insecticides against mites and aphids in apple orchards; psylla in pear orchards; whiteflies, thrips, and mites on vegetables; and whiteflies on cotton. However, sugar esters are not effective against insect eggs. Insecticidal properties of sugar esters were first investigated a decade ago when a scientist noticed that tobacco leaf hairs exuded sugar esters for defense against some soft-bodied insect pests. Similar to insecticidal soap in their action, these chemicals act as contact insecticides and degrade into environmentally benign sugars and fatty acids after application. AVA Chemical Ventures of Portsmouth, NH hopes to have a product based on sucrose octanoate commercially available by the end of 2001.

Because pest resistance to chemical controls has become so common, susceptibility to pesticides is increasingly being viewed by growers as a trait worth preserving. One example of the economic impact of resistance to insecticides has been documented in Michigan, where insecticide resistance in Colorado potato beetle was first reported in 1984 and caused severe economic problems beginning in 1991. In 1991 and following years, control costs were as high as $412/hectare in districts most seriously affected, in contrast to $35-74/hectare in areas where resistance was not a problem. The less a product is applied, the longer a pest population will remain susceptible to that product. Routine use of any pesticide is a problematic strategy.

Botanical pesticides are prepared in various ways. They can be as simple as pureed plant leaves, extracts of plant parts, or chemicals purified from plants. Pyrethrum, neem formulations, and rotenone are examples of botanicals. Some botanicals are broad-spectrum pesticides. Others, like ryania, are very specific. Botanicals are generally less harmful in the environment than synthetic pesticides because they degrade quickly, but they can be just as deadly to beneficials as synthetic pesticides. However, they are less hazardous to transport and in some cases can be formulated on-farm. The manufacture of botanicals generally results in fewer toxic by-products.

Compost teas are most commonly used for foliar disease control and applied as foliar nutrient sprays. The idea underlying the use of compost teas is that a solution of beneficial microbes and some nutrients is created, then applied to plants to increase the diversity of organisms on leaf surfaces. This diversity competes with pathogenic organisms, making it more difficult for them to become established and infect the plant.

An important consideration when using compost teas is that high-quality, well-aged compost be used, to avoid contamination of plant parts by animal pathogens found in manures that may be a component of the compost. There are different techniques for creating compost tea. The compost can be immersed in the water, or the water can be circulated through the compost. An effort should be made to maintain an aerobic environment in the compost/water mixture.

PESTICIDE APPLICATION TECHNIQUES

As monetary and environmental costs of chemical pesticides escalate, it makes sense to increase the efficiency of chemical applications. Correct nozzle placement, nozzle type, and nozzle pressure are very important considerations. Misdirected sprays, inappropriate nozzle size, or worn nozzles will ultimately cost the grower money and increase the risk of environmental damage.

If the monitoring program indicates that the pest outbreak is isolated to a particular location, spot treatment of only the infested area will not only save time and money, but will conserve natural enemies located in other parts of the field. The grower should also time treatments to be least disruptive of other organisms. This is yet another example where knowledge about the agroecosystem is important.

With the increasing popularity of no-till and related conservation tillage practices, herbicide use has increased. One way to increase application efficiency and decrease costs of herbicide use is through band application. This puts the herbicide only where it is needed, usually in soil disturbed by tillage or seed planting, where weeds are most likely to sprout.

Baits and microencapsulation of pesticides are promising technologies. For example, Slam™ is an insecticide-bait mixture for control of corn rootworm. It is a formulation of a bait, curcubitacin B, and carbaryl (Sevin™) in microspheres. It is selective, and reduces the amount of carbaryl needed to control the rootworm by up to 90%. (Remember that crop rotation will generally eliminate the need for any corn rootworm chemical control.)

Another example of bait-insecticide technology is the boll weevil bait tube. It lures the boll weevil using a synthetic sex pheromone. Each tube contains about 20 grams of malathion, which kills the boll weevil. This technique reduces the pesticide used in cotton fields by up to 80% and conserves beneficials. It is most effective in managing low, early-season populations of the boll weevil.

INTEGRATED WEED MANAGEMENT SYSTEMS

Weeds as competitors in crops present a number of unique challenges that need to be recognised when developing management strategies. The intensity of weed problems during a growing season will be influenced by weed population levels in previous years. The axiom "one year's seeding equals seven years' weeding" is apt.

Weed control costs cannot necessarily be calculated against the current year's crop production costs. Weeds present a physical problem for harvesting. Noxious weed seed mixed with grain reduces the price paid to growers. If the seed is sold for crop production the weed can be spread to new areas. For example, the perennial pepperweed, thought to have been introduced to California in sugar beet seed, now infests thousands of acres in the state. In addition, weed economic thresholds must take into account multiple species and variable competetive ability of different crops. For example, 12.7 cocklebur plants in 10 sq. meters of corn cause a 10% yield loss. Only 2 cockleburs in the same area planted to soybeans will cause the same 10% crop loss.

Tactics that can be integrated into weed management systems include:

- *Prevention*—The backbone of any successful weed management strategy is prevention. It is important to prevent the introduction of seeds into the field through sources like irrigation water or manure.
- *Crop rotation*—A practical and effective method of weed management.
- *Cultivation*—Steel in the Field: A Farmer's Guide to Weed Management Tools shows how today's implements and techniques can handle weeds while reducing or eliminating herbicides.
- *Flame weeding*—good for control of small weeds.
- *Delayed planting*—Early-germinating weeds can be destroyed by tillage. And with warmer weather, the subsequently planted crop (depending on the crop, of course) will grow more quickly, thus competing better with weeds.
- *Staggered planting schedule*—This will allow more time for mechanical weed control, if needed. This also lessens the weather risks and spaces out the work load at harvest time.
- *Surface residue management*—As mentioned earlier, a thick mulch may shade the soil enough to keep weed seeds from germinating. In addition, some plant residues are allelopathic, releasing compounds that naturally suppress seed germination.
- *Altered plant spacing or row width*—An example is narrow-row (7–18" between rows compared to conventional 36–39" between rows) soybean plantings. The faster the leaves shade the ground, the less weeds will be a problem.
- *Herbivore*—Cattle, geese, goats, and insects can be used to reduce populations of specific weeds in special situations. Cattle, for example, relish Johnson grass. Weeder geese were commonly used in cotton fields before the advent of herbicides. Musk thistle populations can be satisfactorily reduced by crown- and seed-eating weevils. Goats may be used for large stands of various noxious weeds.
- *Adjusting herbicide use to situation*—Herbicide selection and rate can be adjusted depending upon weed size, weed species, and soil moisture. Young weeds are more susceptible to chemicals than older weeds.

CURRENT STATUS OF IPM

In the last twenty years or so, IPM programs have been developed for important pests in corn, soybeans, cotton, citrus, apples, grapes, walnuts, strawberries, alfalfa, pecans, and most other major crops. These programs are constantly being revised or fine-tuned, and occasionally undergo a significant overhaul as the introduction of a new technology or new pest makes the present IPM program obsolete.

The best source of information on conventional IPM is the Cooperative Extension Service (CES) associated with the land-grant university in each state. Booklets and fact sheets describing IPM programs and control measures for a wide range of crops and livestock are available free or for a small charge. In 1993, leaders from USDA, EPA, and FDA announced a goal of placing 75% of U.S. crop acreage under IPM by the year 2000. The IPM Initiative described three phases:

1. Create teams of researchers, Extension personnel, and growers to propose projects to achieve the 75% goal.
2. Fund the best of those projects.
3. Facilitate privatisation of IPM practices developed in the process.

Although some progress is evident, the Initiative has not received full funding from Congress. In addition, the USDA's criteria for measurement have been criticised for not distinguishing between practices that are related to "treatment" and those that are "preventive," that is, based on altering the biological and ecological interactions between crops, pests, and beneficial organisms. Practices that constitute "treatment" with or contribute to the efficiency of pesticides are considered as "indicative of an IPM approach" by USDA's criteria, as are practices that draw upon and are most compatible with biological relationships on the farm.

The primary goal of biointensive IPM is to provide guidelines and options for the effective management of pests and beneficial organisms in an ecological context. This requires a somewhat different set of knowledge from that which supports conventional IPM, which in turn requires a shift in research focus and approach. Recommended actions to better facilitate the transition to biointensive IPM are:

- Build the knowledge/information infrastructure by making changes in research and education priorities in order to emphasize ecology-based pest management.
- Redesign government programs to promote biointensive IPM, not "Integrated Pesticide Management."
- Offer consumers more choices in the marketplace.
- Use the market clout of government and large corporations.
- Use regulation more consciously, intelligently, and efficiently.

FUTURE OF IPM

IPM in the future will emphasize biological and ecological knowledge in managing pests. Beyond that, specific areas are described here that will impact research and implementation of IPM in the future.

The FQPA, the amended Federal Insecticide, Fungicide, and Rodenticide Act (FIFRA), requires the EPA to review all federally registered pesticides in the next 10 years and to use a more comprehensive health standard when allowing re-registration. The ultimate impact is unknown, but FQPA will most likely result in stricter regulations concerning pesticide residues in food, particularly with respect to organochlorines, organophosphates, and carbamates. Some of the most toxic pesticides have already been "de-registered" with respect to some of their former uses. These regulations may provide incentive for more widespread adoption of IPM.

Pest control methods are evolving and diversifying in response to public awareness of environmental and health impacts of synthetic chemical pesticides and resulting legislation. The strong growth of the organic foods market—20% annual expansion for the past several years—may also be a factor in the accelerated development of organic pest management methods.

Agricultural pests are developing resistance to many synthetic agrichemicals, and new synthetic chemicals are being registered at a slower rate than in the past. This situation has helped open the market for a new generation of microbial pesticides.

Research is proceeding on natural endophytes—fungi or bacteria that have a symbiotic (mutually beneficial) relationship with their host plant—and their effects on plant pests. This research might yield products that could be used to inoculate plants against certain pests. Synthetic beneficial attractants such as Predfeed IPM™ and L-tryptophan may help increase the efficacy of natural controls by attracting beneficials to a crop in greater numbers than usual.

Weed IPM

Weeds are the major deterrent to the development of more sustainable agricultural systems, particularly in agronomic crops. Problems associated with soil erosion and water quality are generally the result of weed control measures like tillage, herbicides, cultivation, planting date and pattern, etc. In the future, research will focus not on symptoms, such as soil erosion, but on basic problems such as how to sustainably manage soils. Weeds, as an important facet of sustainable soil management, will consequently receive more emphasis in IPM or Integrated Crop Management (ICM) programs.

On-farm Resources

As farm management strategies become increasingly fine-tuned to preserve a profitable bottom line, the conservation, utilisation, and development of on-farm resources will take on added importance. In the context of IPM, this will mean greater emphasis on soil management as well as on conserving beneficial organisms, retaining and developing beneficial habitats, and perhaps developing on-farm insectaries for rearing beneficial insects.

IPM Online

There is an increasing body of information about production, marketing, and record-keeping available to growers via the Internet. The Internet is also a good source of information about IPM, beneficial insects, products, and pest control options for individual crops. IPM specialists are generating high-quality Web sites as a modern educational delivery tool, and many Extension Service leaflets are now being made available in electronic format only. This trend will only accelerate as more and more agriculturists familiarise themselves with the Internet.

IPM Certification and Marketing

Certification of crops raised according to IPM or some other ecology-based standards may give growers a marketing advantage as public concerns about health and environmental safety increase. For example, since 1995, Wegmans has sold IPM-labeled fresh-market sweet corn in its Corning, Geneva, Ithaca, Syracuse, and Rochester, New York stores.

Wegmans has also added IPM-labeled corn, beets, and beans to its shelves of canned vegetables. One goal of the program, in addition to being a marketing vehicle, is to educate consumers about agriculture and the food system. Another goal is to keep all growers moving along the "IPM Continuum." Growers must have an 80% "score" on the IPM program elements within three years, or face losing Wegmans as a buyer. These "ecolabels," as they're known, are becoming more popular, with over a dozen brands now in existence. They may provide for a more certain market and perhaps a price premium to help growers offset any costs associated with implementing sustainable farming practices.

A possible downside to implementing such programs is that they require additional paperwork, development of standards and guidelines, and inspections. There is concern from some quarters that IPM labeling will cause consumers to raise more questions about pesticide use and the safety of conventional produce. Some advocates of organic farming worry about consumer confusion over the relationship of the ecolabel to the

"Certified Organic" label. Mothers & Others for a Livable Planet, a national, non-profit, consumer advocacy and environmental education organisation, has partnered with apple farmers in the Northeast region to create a supportive market environment for farm products that are locally grown and ecologically responsible. The result is the Core Values eco-label: A CORE Values Northeast apple is locally grown in the Northeast by farmers who are striving to provide apples of superior taste and quality while maintaining healthy, ecologically balanced growing environments. Growers whose apples bear the CORE Values Northeast seal are accredited in knowledge-based biointensive Integrated Pest Management (IPM) production methods.

The Homegrown Wisconsin ecolabel is a result of a collaboration between the World Wildlife Fund (WWF), the Wisconsin Potato and Vegetable Growers Association (WPVGA), and the University of Wisconsin. Raising consumer demand for biology-based-IPM farm products is the goal of the program. There has been an IPM labeling program casualty in 2000. Massachusetts's "Partners with Nature" marketing program closed its doors after losing funding support from the Massachusetts Department of Food and Agriculture. The program, which included IPM production guidelines, had operated since 1994, with 51 growers participating in 1999.

REFERENCES

Bennett, Gary W., Ph.d., Owens, John M., Ph.d., Corrigan, Robert M, Ph.d. *Truman's Scientific Guide to Pest Management Operations,* 6th Edition, pages 10, 11, 12, Purdue University, Questex, 2005.

Jahn, GC, JA Litsinger, Y Chen and A Barrion. 2007. *Integrated Pest Management of Rice: Ecological Concepts.* In Ecologically Based Integrated Pest Management (eds. O. Koul and G.W. Cuperus). CAB International Pp. 315-366.

Kogan, M 1998. "Integrated Pest Management:Historical Perspectives and Contemporary Developments", *Annual Review of Entomology* Vol. 43: 243-270.

12

Fumigation for Soil Pest Control

Fumigation continues to play a valuable role in many pest control operations; however, both the concepts and the procedures for controlling insects and other organisms are changing. With increased public concern over the adverse effects of pesticidal chemicals on human health and the environment, greater emphasis is being given to methods that can circumvent the use of these materials. Nevertheless, the need for chemical pesticides, particularly the fumigants, is likely to continue for many years to come; fumigants have unique properties and capabilities that permit use in numerous situations where other forms of control are not feasible or practical.

Fumigants are a unique and particularly valuable group of pesticides that can kill insects where no other form of control is feasible. To a large extent they are irreplaceable. The use of certain fumigants has been restricted in some countries because of suspected adverse effects. Excessive use of fumigants or the misuse of them to cause accidents and produce adverse publicity is likely to bring about even greater restrictions in their use. By careful planning and management, fumigation may be incorporated into food preservation systems so that fumigants can be used more effectively and safely than when used independently. They should never be used as a substitute for sound management and good sanitation procedures. The benefits derived can include reduced cost of storage with improved food quality, reduced residues in food materials, greater occupational safety and less environmental contamination. All of these benefits are of great concern to the general public and will be factors that have to be taken into consideration in the future use of fumigants. The ultimate goal in the control of pests in stored products should be to so improve the methods of handling, storing and processing commodities, that the need for pesticides will decrease. Fumigants will then only be needed when unavoidable infestations are encountered.

CHOICE OF FUMIGANT

There are many chemical compounds which are volatile at ordinary temperatures and sufficiently toxic to fall within the defintion of fumigants. In actual practice, however, most gases have been eliminated owing to unfavourable properties, the most important being chemical instability and destructive effects on materials. Damage to materials may take place in several ways, as follows:

1. Excessively corrosive compounds attack shipping containers or spoil the structure and fittings of fumigation chambers or other spaces undergoing treatment.
2. Reactive chemicals form irreversible compounds, which remain as undesirable residues in products. In foodstuffs such reactions may lead to taint or the formation of poisonous residues. Other materials may be rendered unfit by visible staining or by the production of unpleasant odours.
3. Physiologically active compounds may destroy or severely injure growing plants, fruit or vegetables, and may adversely affect seed germination.

Highly flammable compounds are not necessarily excluded if dangers of fire and explosion can be controlled by the addition of other suitable compounds, or if fumigation procedures are carefully designed to eliminate these hazards. Toxicity to human beings is not necessarily a cause for exclusion. All known fumigants are toxic to humans to a greater or lesser degree and ways can be devised for their safe handling under the required conditions of application. However, some commonly used compounds have been shown to be capable of producing long-term effects that were previously unknown. The use of such fumigants is becoming more restricted and some materials have already been eliminated from the list of fumigants approved for use in certain countries.

EVAPORATION OF FUMIGANTS

Boiling Point

The boiling point of different chemical compounds generally rises with the increase of molecular weights. If the highest possible concentrations are required at the beginning of the fumigation with such compounds, more rapid volatilsation will have to be effected in some way.

From the physical standpoint, fumigants may be divided into two main groups according to whether they boil above or below room or moderate outdoor temperatures (20°C to 25°C). The low boiling point fumigants, such as methyl bromide, may be referred to as gaseous -type fumigants. These are kept in cylinders or cans designed to withstand the pressure exerted by the gas at the highest indoor or outdoor temperatures likely to be encountered. The second main group of fumigants contains those with high boiling points; these are usually described as liquid-type or solidtype according to the form in which they are shipped and handled. In some kinds of work, such as grain and soil fumigation, the slow evaporation of certain liquids is an advantage because the initial flow leads to a better distribution of the gas subsequently volatilsed. In other applications, where personnel have to distribute the fumigants, slow evaporation of the liquids or solids makes them safer to handle.

Included in the general term solid-type fumigants are certain materials which are not fumigants themselves, but which react to form fumigants after application. Examples are calcium cyanide powder, which reacts with atmospheric moisture to yield hydrogen cyanide (HCN), and formulations of aluminium and magnesium phosphides which also react with moisture to produce phosphine (hydrogen phosphide).

There are also some fumigants in the form of crystals and flakes that sublime to give off fumigant vapours. Examples are paradichlorobenzene and naphthalene.

Maximum Concentrations

The maximum weight of a chemical that can exist as a gas in a given space is dependent on the molecular weight of that chemical. This fact, implicit in the well-known hypothesis of Avogadro, has an important practical application. It is useless attempting to volatilise in an empty chamber more fumigant than can exist in the vapour form.

Commodities treated and remarks

Tobacco and plant products; also spot treatment. Injures growing plants, fresh fruit and vegetables. Marketed with carbon tetrachloride.

Grain. Usually as ingredient of nonflammable mixtures. Only weakly insecticidal. Used chiefly in mixture with flammable compounds in grain fumigation to reduce fire hazard and aid distribution.

Grains and plant products. Injurious to living plants, fruit and vegetables. Highly irritating lachrymator. Bactericidal and fungicidal.

Insects in open space of structures. Does not penetrate commodities.

General fumigant. Particularly useful for certain fruit; may injure growing plants.

Seeds and grains. Usually mixed with carbon tetrachloride.

Grains, cereals and certain plant products. Toxic at practical concentrations to many bacteria, fungi and viruses. Strongly phytotoxic and affects seed germination.

Application to individual packages of dried fruit.

General fumigant, but may be phytotoxic. Safe on seeds but not recommended for fresh fruit and vegetables.

General fumigant. May be used with caution for nursery stock, growing plants, some fruit and seeds of low moisture content.

Usually mixed with CO_2. Formerly used for grain, now mainly for stored furs.

Control borers in peach trees and soil insects. Applied as crystals. May affect seed germination.

Grain and processed food fumigant; gas generated from aluminium or magnesium phosphide.

Control of dry-wood termites in structures.

Nonflammable ingredient of grain fumigants. Sometimes used alone.

Latent Heat of Vaporisation

Unless it is sustained by warming from an outside source, the temperature of an evaporating liquid constantly drops owing to the fall in energy caused by the escape of molecules with greater than average energy. Thus, evaporation takes place at the expense of the total heat energy of the liquid. The number of calories lost in the formation of one gramme of vapour is called the latent heat of vaporisation of the liquid. Some fumigants have higher latent heats than others.

Both HEN and ethylene oxide, with latent heats of 210 and 139 respectively, absorb considerably more heat in passing from liquid to than do methyl bromide and ethylene dibromide, with latent heats of 61 and 46 respectively.

The factor of latent heat is of important practical significance. The high pressure fumigants, such as HCN, ethylene oxide and methyl bromide, are usually kept under pressure in suitable cylinders or cans. On release into the atmosphere, volatilisation takes place rapidly and, unless the lost heat is restored, the temperature of the fumigant may fall below the boiling point and gas may cease to be evolved. Also, as the liquid changing to gas is led through metal pipes and tubes, or rubber tubing, the fall in temperature may freeze the fumigant in the lines and prevent its further passage. In many applications, to be described elsewhere in this manual, it is advisable to apply heat to the fumigant as it passes from the container into the fumigation space.

Fumigants that are liquids at normal temperatures and are volatilised from evaporating pans or vaporising nozzles may require a source of heat, such as a hot plate, in order that full concentrations may be achieved rapidly.

Diffusion and Penetration

Fumigants are used because they can form insecticidal concentrations: (a) within open structures or (b) inside commodities and in cracks and crevices into which other insecticides penetrate with difficulty or not at all. Hence, it is necessary to study the factors that influence the diffusion of gases in every part of a fumigation system. This study includes the behaviour of fumigants both in empty spaces and also in structures loaded with materials into which the gas is required to penetrate.

Law of diffusion

Graham's law of diffusion of gases states that the velocity of diffusion of a gas is inversely proportional to the square root of its density. Also, the densities of gases are proportional to their molecular weights. Therefore, a heavier gas, such as ethylene dibromide, will diffuse more slowly throughout an open space than a lighter one such as ethylene oxide. While this basic law is of importance, especially for empty space fumigations, the movement of gases in contact with any internal surface of the structure or within any contained materials is greatly modified by the factor of sorption discussed below. The rate of diffusion is also directly related to temperature, so that a given gas will diffuse more quickly in hot air than in cold air.

Specific gravity and distribution

Many of the commonly used fumigants are heavier than air. A notable exception is hydrogen cyanide. If a gas heavier than air is introduced into a chamber filled with air and it is not agitated by fans or other means, it will sink to the bottom and form a layer below the air. The rate of mixing between the two layers may be very slow. For example, in a fumigation of the empty hold of a ship with the heavy gas methyl bromide where the fumigator had neglected to place a circulating fan, a sharp demarcation was observed between the lower half with the gas, where all of the insects were killed, and the upper part, where complete survival occurred.

In good fumigation practice, settling or stratification will not be encountered if adequate provision is made to disperse the gas properly from the very beginning of the treatment. Even distribution can be ensured by employing singly or in suitable combination:multiple gas inlets, fans or blowers and/or circulation by means of ducts and pipes. Contrary to popular belief, once a gas or number of gases heavier than air have been thoroughly mixed with the air in a space, settling out or stratification of the heavier components takes place very slowly; so slowly, in fact, that once a proper mixture with air has been secured, the problem of stratification of a heavierthan-air fumigant is of no practical importance for the exposure periods commonly used in fumigation work.

Mechanical aids to diffusion

It has already been suggested that distribution and penetration can be aided and hastened by the use of blowers and fans. Such propellers may work free in the structure or through a system of circulating ducts. These devices may also add greatly to the efficiency of the fumigation process by hastening the volatilisation of high boiling point liquids from evaporating pans and by preventing stratification of heavy gases. Also, a factor known as the Turtle effect* has proved useful in the fumigation of certain materials susceptible to injury. It was shown that rapid stirring by a centrifugal fan in a fumigation chamber at atmospheric pressure greatly hastened the attainment of uniform concentrations of methyl bromide in all parts of a load of early potatoes, so that the consignment was not overdosed at the outside of the packages or under dosed at the centre. In a four-hour exposure period, rapid stirring for one hour at the beginning of the treatment was, to all intents and purposes, as effective as continuous stirring for the whole time.

Sorption

A very important factor affecting the action of fumigants is the phenomenon known as sorption. It is not possible in this manual to give a complete explanation of sorption, because the interaction of all forces involved is complex. Fortunately, for the purpose of understanding fumigation practice, it is possible to give a general account of the important factors concerned. In the relationship of gases to solids, sorption is the term used to describe the total uptake of gas resulting from the attraction and retention of the molecules by any solid material present in the system. Such action removes some of the molecules of the gas from the free space so that they are no longer able to diffuse freely throughout the system or to penetrate further into the interstices of the material. In fumigation practices, collision with air molecules tends to slow down gaseous diffusion through the material and sorption takes place gradually. Thus, there is a progressive rather than immediate lowering of the concentrations of the gas in the free space.

The curves for each of the four compounds clearly show the differences in degree of sorption of the fumigants by the same load in the chamber. Throughout the exposure period of six hours, the fall in concentration of methyl bromide was proportionately less compared with that for the three other fumigants, both in the empty chamber and with the two loads of oranges. This was due to the fact that the internal surface of the chambers and the boxes of oranges both sorbed less of the methyl bromide than of the other gases in proportion to the applied dosage. Sorption under a given set of conditions determines the dosage to be applied, because the amount of fumigant used must be sufficient both to satisfy the total sorption during treatment and also to leave enough free gas to kill the pest organisms. The general term sorption covers the phenomena of adsorption and absorption. These two are reversible because the forces involved, often referred to as van der Waal's forces, are weak. On the other hand, a stronger bonding called chemisorption usually results in chemical reaction between the gas and the material and is irreversible under ordinary circumstances.

Physical sorption

From the point of view of practical fumigation, adsorption and absorption, being both physical in nature and reversible, may be discussed in this manual under the heading of physical sorption. However, it is necessary to make some distinction between them at the outset because the forces involved may be less with adsorption than with absorption. Stated briefly, adsorption is said to occur when molecules of a gas remain attached to the surface of a material. Because some absorbents, such as charcoal or bone meal, are highly porous bodies with large internal surfaces, adsorption may also occur inside a given body.

Absorption occurs when the gas enters the solid or liquid phase and is held by capillary forces that govern the properties of solutions. For instance, a gas may be absorbed in the aqueous phase of grain or in the lipid phase of nuts, cheese or other fatty foods.

Physical sorption, considered generally, is an extremely important Factor affecting the successful outcome of fumigations. Apart from specific reactions between certain gases and commodities, it may be stated as a general rule that those fumigants with higher boiling points tend to be more highly sorbed than the more volatile compounds.

Physical sorption varies inversely as the temperature, and is thus greater at lower temperatures. This fact has important practical applications. It is one of the reasons why dosages have to be progressively increased as the temperature of fumigation is lowered.

Sorption may also be influenced by the moisture content of the commodity being fumigated. This was demonstrated by Lindgren and Vincent in the fumigation of a number of foodstuffs with methyl bromide; at higher moisture contents more fumigant was sorbed. This effect may be important with fumigants which are soluble in water to any significant degree.

The specific physical reaction between a given gas and a given commodity cannot be accurately predicted from known laws and generalisations. Usually, a certain fumigant must be tested with each material concerned before a recommendation for treatment can be drawn up.

Desorption

When a treatment is completed and the system is ventilated to remove the fumigant from the space and the material, the fumigant slowly diffuses from the material. This process is called Resorption and is the reverse of physical sorption. With the common fumigants and the commodities usually treated, residual vapours are completely dissipated within reasonable periods, although the length of time varies considerably according to the gas used and the material treated. Because of the inverse effect of temperature, dissipation of the fumigant usually takes place more slowly when the material is cold and may be hastened by warming the space and its contents.

Humidity also facilitates desorption of fumigants; at high humidity, wheat fumigated with ethylene dibromide was found to desorb 80 percent more of the fumigant than at very low humidity. As humidities can change appreciably with changing temperature, the rate of desorption may be dependent on the combined effect of both factors. Removal of desorbing gas can be speeded up by employing fans and blowers to force fresh air through the material. Natural ventilation may be hastened by taking the goods out of doors where advantage can be taken of wind, thermal air currents and the warming effect of sunlight. Some of the residual fumigant, usually small in quantity, may not be desorbed because of chemical reaction with the material.

Chemical reaction

If chemical reaction takes place between the gas and the material, new compounds are formed. This reaction is usually characterised by specificity and irreversibility. If the reaction is irreversible, permanent residues are formed. Examples are the reaction between hydrogen cyanide (HCN) and the reducing sugars in dried fruits with the formation of cyanohydrins or the appearance of inorganic bromide compounds after treatment of some foodstuffs with methyl bromide.

Because this type of reaction is essentially chemical it may be expected that its intensity varies directly with the temperature. This assumption has been confirmed by observation. Dumas has reported proportionately less fixed bromide residues in fruits as the temperature of fumigation was reduced from 25 to 4°C. Lindgren et al found an increase in the bromide content of wheat as the temperature during fumigation rose from 10 to 32°C.

Residue tolerances

In recent years attention has been focussed on the nature and possible effects on human beings of insecticidal residues appearing in foodstuffs. World-wide interest in this problem is reflected in the fact that international organisations such as the Food and

Agriculture Organisation of the United Nations (FAO) and the World Health Organisation (WHO) have set up special committees to investigate and report on the nature and significance of residues formed in foodstuffs as the result of the application of pesticides at different stages (as seed dressings, during growth, storage, transportation, etc.) prior to human consumption. These special committees review a number of pertinent factors involved in the use of each pesticide. Important factors, among others, are the toxicological significance of any residues formed and the average fraction of the total diet likely to be constituted by a food containing this residue. Through their Codex Alimentarius Committee these organisations undertake "to recommend international tolerances for pesticide residues in specific food products." Such recommendations are not binding on Member Nations of these organisations but are intended to be used as guides when particular countries are formulating their own regulations for pesticide residue tolerances.

Other Effects on Materials

Apart from the question of significant residues in foodstuffs, there is the problem of other effects which have a direct bearing either on the choice of the particular fumigant or on the decision as to whether fumigation is possible at all. The main types of reaction may be summarised as follows:

Physiological effects

1. Nursery Stock and Living Plants
   - (a) Stimulation of growth
   - (b) Retardation of growth
   - (c) Temporary injury and subsequent recovery
   - (d) Permanent injury, usually followed by death
2. Seeds
   - (a) Stimulation of germination
   - (b) Impairment or total loss of germination
   - (c) Poor growth of seedlings from germinated seeds
3. Fruit and Venetables
   - (a) Visible lesions
   - (b) Internal injury

(c) Shortening of storage life

(d) Delay of ripening

(e) Stimulation of storage disorders

4. Infesting Organisms

(a) Death

(b) Stimulation of growth or metamorphosis

(c) Delay in development

(d) Stimulation of symptoms of disease (so-called "diagnostic effect")

Physical and chemical effects on nonliving materials

1. Production of foul or unpleasant odours in furnishings or materials stored in premises.
2. Chemical effects that spoil certain products (for example, some fumigants render photographic films and papers unusable).
3. Reaction with lubricants followed by stoppage of machinery (clocks will often stop after fumigation with HCN).
4. Corrosive effects on metals (phosphine reacts with copper, particularly in humid conditions).

Dosages and concentrations

There should be a clear understanding of the difference between dosage and concentration. The dosage is the amount of fumigant applied and is usually expressed as weight of the chemical per volume of space treated. In grain treatments, liquid-type fumigants are often used and the dosage may be expressed as volume of liquid (litres or gallons) to a given volume (amount of grain given as litres or bushels) or sometimes to a given weight (quintals, metric tonnes or tons).

From the moment that a given dosage enters the structure being fumigated, molecules of gas are progressively lost from the free space either by the process of sorption and solution described above or by actual leakage from the system, if this occurs. The concentration is the actual amount of fumigant present in the air space in any selected part of the fumigation system at any given time. The concentration is usually determined by taking samples from required points and analysing them. It may thus be said that the dosage is always known because it is a pre-determined quantity. Concentration has to be determined because it varies in time and position according to the many modifying factors encountered in fumigation work. Three methods of expressing gas concentrations in air are in common use: weight per volume, parts by volume and percent by volume.

Weight per volume

For practical designation of dosages, this is the most convenient method because both factors—the weight of the fumigant and the volume of the space—can be easily determined. In countries using the metric system, this is usually expressed in grammes per cubic metre (g/m^3), whereas in countries using the British system of weights and measures, expression is usually in terms of pounds or ounces avoirdupois (avdp) per 1 000 cubic feet (lb/1 000 ft^3 or oz/1 000ft). By a fortunate coincidence in units of measurement, grammes per cubic metro are, for all practical purposes, equal to ounces per thousand cubic feet. Thus, recommended dosages can readily be converted from one system to the other.

Parts or percent by volume

Parts by volume and percent by volume will be discussed together because both modes of expression give the relative numbers of molecules of gas present in a given volume of air. The values for both modes have the same digits, but the decimal points are in different places (3 475 parts per million by volume of a gas is the same as 0.3475 percent by volume). Parts per million of gases in air are used in human and mammalian toxicology and in applied industrial hygiene. Percent by volume is used in expressing the flammability and explosive limits of gases in air.

Calculations for conversion of concentration values

By means of simple calculations giving useful approximations, values may be converted from weight per volume to parts by volume and vice versa. These calculations take into account the molecular weight of the gas and the fact that, with all gases, the gramme molecular weight of the substance occupies 22.414 litres at 0°C and 760 millimetres pressure.

A. To convert grammes per cubic metre (or milligrammes per litre or ounces per 1 000 cubic feet) into parts by volume.

1. Divide the given value by the molecular weight of the gas and multiply by 22.4; the resulting figure is the number of cubic centimetres (cm) of gas per litre of air.
2. One thousand times the figure obtained is the value in parts per million by volume.

3. One tenth of the figure obtained in (1) is the percentage by volume.

B. To convert parts per million (or percentage of volume) of gases to grammes per cubic metre (or milligrammes per litre or ounces per 1 000 cubic feet):

1. Divide the parts per million by 1 000, or multiply the percentage by ten to give the number of cubic centimetres of gas per litre of air.
2. Multiply this figure by the molecular weight of the gas in question and divide by 22.4. Comparative figures for weights and volumes at various levels have been calculated for the important gases, and these are given in the tables accompanying the subsequent discussion of each particular gas.

Concentration X time (c x t) Products

Most fumigation treatments are recommended on the basis of a dosage given as the weight of chemical required for a certain space—expressed as grammes per cubic metre or pounds per 1 000 cubic feet or as volume of liquid applied to a certain weight of material—expressed as litres per quintet or gallons per 1 000 bushels. Usually, this designation of dosage is followed by a statement of the length of the treatment in hours and the temperature or range of temperature at which the schedule will apply.

While such recommendations are usually based on treatments that have proved successful under certain conditions, they should also take into account the fact that certain factors may modify the concentrations left free to act against the insects. One important factor already mentioned is the effect of loads of different sizes. Another is the leakage from the structure undergoing treatment. What is really important is the amount of gas acting on the insects over a certain period of time. For instance, it is known (Bond and Monro, 1961) that in order to kill 99 percent of larvae of Tenebroides mauritanicus (L.) at 20°C, a concentration of 33.2 milligrammes per litre of methyl bromide must be maintained for 5 hours.

The product 33.2 milligrammes per litre x 5 hours = 166 milligrammes per litre x hours is known as the concentration x time product needed to obtain 99 percent control of this insect. It can be abbreviated and referred to as the c x t product. In the literature it is often expressed numerically with the notation mg h/l (milligramme hours per litre). In this example it would be known as the lethal dose for 99 percent of the population, or the LD.

In order to apply this method of treatment designation to practical fumigations, it is necessary to make reasonably correct determinations of the fumigant concentrations required to kill the insects under certain specific conditions; important modifying conditions are temperature and humidity.

Table 1. Required concentration X Time (c x t) products to obtain 99 percent mortality of tenebroides mauritanicus

Concentration methyl bromide	*Exposure*	*c x t product*
mg/l		*hours mg h/l*
83	2	166
55.3	3	166
41.5	4	166
33.2	5	166
23.7	7	166
16.6	10	166

It must be emphasized again that before they are applied in practical use each product must be calculated for the different stages of an insect species at a certain temperature and humidity. Under practical conditions, variations in temperature are particularly important. In practice, several insect species or stages of a given insect may be treated and therefore the c x t product required is that which is effective against the most tolerant species or stage present in the system.

Table 2. Integrated concentration X time products within the infested commodity

Hours	*Rectangle*	*Triangle*	*Total area*	*Cumulative*
		mg h/l		
1		3	3	3
2	6	2.9	8.9	11.9
3	11.7	1.65	13.35	25.25
4	15	1	16	41.25
5	17	0.5	17.5	58.75
6	18	0.5	18.5	77.25
7	19	0.25	19.25	96.5
8	20	-	20	116.5
9	20	-	20	136.5
10	20	-	20	156.5
11	20	-	20	176.5
11.7	13.5	-	13.5	190.0

The value and possible application of the c x t product for the fumigation of insects has been investigated by a number of workers. The important modifying effects of temperature, humidity and the moisture content of the commodity are emphasized. Kenaga described the use of graphs to estimate the effective use of c x t products of eight different fumigants against Tribolium confusum Duv. under varying conditions of time and temperature. Heseltine and Royce showed how integrated c x t products of ethylene oxide and methyl bromide may be used in practice with the aid of specifically designed concentration indicators in the form of sachets. The use of integrated c x t products is particularly useful in routine fumigations when the reaction of a particular species or groups of species has been carefully worked out under the range of conditions likely to be encountered. It has been used successfully in large-scale eradication campaigns.

Table 3. Response of Methyl Bromide-resistant granary weevil to six other fumigants

Fumigant	*Resistant*	*Normal*	*Tolerance ratio*
Methyl bromide	19.7	3.6	5.5
HCN	16.4	8.2	2.0
Acrylonitrile	4.9	1.05	4.7
Ethylene oxide	20.1	4.1	4.8
Chloropicrin	6.6	3.9	1.7
Phosphine	13.0	2.2	5.9
Ethylene dibromide	8.5	2.85	3.0

Dosage schedules are, perhaps, best given in terms of weight of chemical required for a certain space for a specified period of time along with the c x t products necessary to achieve control. Thus by monitoring gas concentrations during treatment, an applicator can add gas, extend the exposure or make other changes necessary to ensure success. For plant quarantine work, recommendations based on the c x t principle are particularly valuable because they promote uniformity in standards and permit reliable certification of goods so treated. Schedules based on these concepts are in use in several countries, e.g. Plant Protection and Quarantine Treatment Manual. For other treatments of stored products, where sorption in the commodity is appreciable, schedules based on the c x t principle but given in terms of weight of fumigant per unit volume of space and per unit weight of goods for specified exposure times have been worked out for some commodities.

While the c x t method is useful for most fumigants, it cannot be employed with phosphine. Although concentration and exposure time are still the main factors that determine toxicity of this fumigant, the length of the exposure time is of great importance. Phosphine is a slow acting poison that is absorbed slowly by some insects even at high concentrations. Therefore, high concentrations may not increase toxicity; in fact, they may cause insects to go into a protective narcosis. In a phosphine fumigation certain minimum concentrations are required, and therefore gas analysis should be carried out to ensure the presence of sufficient gas. For most treatments the manufacturers' directions will provide adequate treatment if no excessive loss through leakage or sorption occurs and adequate periods are allowed under gas.

TOXICITY OF FUMIGANTS TO INSECTS

As far as is known at present, fumigants enter the insect mainly by way of the respiratory system. The entrance to this system in larvae, pupae and adults is through the spiracles, which are situated on the lateral surfaces of the body. The opening and closing of the spiracles are under muscular control. To enter insect eggs, gases diffuse through the shell (chorion) of the egg or through specialised "respiratory channels". It has been shown that some gases may diffuse through the integument of insects, but at present the comparative importance of this route for the entry of fumigants is not known. It is known that the poisoning of an insect by a fumigant is influenced by the rate of respiration of that insect; any factor that increases the rate of respiration tends to make the insect more susceptible. The practical significance of the more important factors influencing the toxic action of fumigants is discussed in the following paragraphs.

Effect of Temperatre

General effects

The most important environmental factor influencing the action of fumigants on insects is temperature. In the range of normal fumigating temperatures from 10 to 35°C, the concentration of a fumigant required to kill a given stage of an insect species decreases with the rise in temperature. From the purely biological standpoint, this is mainly due to the increased rate of respiration of the insects in response to the rise in temperature. Physical sorption of the fumigant by the material containing the insects is reduced and proportionately more fumigant is available to attack the insects. Therefore, within the range mentioned, conditions for successful fumigation improve as the temperature rises. These conditions are reflected in the schedules for recommended treatments included in this manual.

Low temperature fumigation

At temperatures below 10°C, the situation is more complicated. Below this point, increased sorption of the gas by the body of the insect may counterbalance the effects of decrease in respiration, and also the resistance of insects may be weakened by the effects of exposure to low temperatures. With some fumigants, less gas is required to kill certain species as the temperature is raised or lowered on either side of some point at which the insects are most tolerant. However, with others, toxicity to the insects declines as the temperature falls; for example, with methyl bromide there is a moderate decrease in toxicity down to the boiling point and below this temperature effectiveness drops off sharply so that the amount of gas required to kill the insects increases dramatically. For the reasons already given in the previous discussion, at lower temperatures sorption of the fumigant by the infested material is increased and more fumigant must be applied to compensate for this. Also, diffusion of a gas is slowed down in relation to reduction in temperature.

Prefumination and postfumioation temperatures

It is important to bear in mind that the results of a fumigation may be influenced not only by the temperature prevailing during the treatment, but also by the temperatures at which the insects are kept before and after treatment.

If the insects have been kept in a cool environment, their metabolic rate will be low. If they are immediately fumigated at a higher temperature, their physiological activity may still be influenced by their previous history, and the uptake of the poison may not be as great as if they had been kept at the temperature of fumigation for a long time previous to treatment. These phenomena can be of practical significance, particularly for certain species of insects that may go into a state known as diapause for description of diapause and list of species involved). For insects in this state, tolerance to some fumigants, e.g. methyl bromide and phosphine, may be several times greater than for non-diapausing insects. For other species not in diapause, toxicity is usually found to be closely dependent on the temperature of the fumigation.

A fumigator must have some knowledge of the previous history of the infested material as well as the species to be treated if he or she is to apply the recommended fumigation treatments most effectively. In all treatments, the material should be warmed to the treatment temperature for several hours to bring the insects to corresponding physiological activity before fumigating. If species disposed to the state of diapause are present the dosage and exposure applied should be increased to a level that will kill the most tolerant insects.

Effect of Humidity

From the present knowledge of insect toxicology, it is not possible to make any general statements about the influence of humidity on the susceptibility of insects to fumigants.

Variations in response at certain humidities have been observed not only between different species subjected to different fumigants but also between stages of the same species exposed to a single fumigant. However, variations due to humidity are not so important in practice as those due to temperature.The treatments recommended in this manual are adequate for the range of moisture content and humidity normally encountered.

Effect of Carbon Dioxide

Carbon dioxide, in certain concentrations, may stimulate the respiratory movements and opening of spiracles in insects. A number of authors have shown that addition of carbon dioxide to some of the fumigants may increase or accelerate the toxic effect of the gas. With each fumigant acting on different insects, there seems to be an optimum amount of carbon dioxide needed to provide the best insecticidal results. Excessive amounts of carbon dioxide tend to exclude oxygen from insects and thus interfere with the action of the fumigants.

With certain fumigants, such as ethylene oxide and methyl formate, the addition of carbon dioxide may work to advantage both by reducing the fire or explosion hazards and by increasing the susceptibility of the insects. On the other hand, with fumigants that are nonflammable, the advantages of adding carbon dioxide may be offset by the extra cost and work required to handle the additional weight of containers.

Protective Narcosis

Some fumigants can produce paralysing effects on insects that may alter the toxicity of these or other fumigants. In the use of hydrogen cyanide (HCN) against insects, it has been shown that, if certain species are exposed to sublethal concentrations before the full concentration is applied, the resulting fumigation is less effective than one in which the insects are subjected to the full concentration from the very beginning. A similar protective effect can also occur with the fumigant phosphine if insects are exposed to excessive concentrations during a treatment. Also, insects that have been narcotised by sublethal concentrations of HCN have been found to be protected from lethal treatments with other fumigants, e.g. methyl bromide and phosphine. This effect has been referred to as "protective stupefaction" or "narcosis".

Although phosphine itself can narcotise insects it does not, however, protect them from the action of methyl bromide as does HCN; in fact, phosphine and methyl bromide can be used together as a "mixture" to enhance the effectiveness of each other. From the practical point of view the phenomenon of narcosis is important because it can reduce the effectiveness of certain fumigants. However, steps can be taken to avoid problems of this nature:

1. In fumigations with HCN the maximum concentration attainable from a recommended dosage should be achieved as soon as possible at the beginning of the treatment.
2. HCN should not be applied with other fumigants such as methyl bromide or phosphine, if the maximum toxic effect is to be achieved.
3. Excessive concentrations of phosphine likely to produce a protective narcosis should not be used.

Fluctuations in Susceptibility of Insects

It has often been observed that there may be fluctuations in the susceptibility of populations of insects to a given poison. Some of the reasons are known, while others need further clarification. Two important factors are undoubtedly seasonal changes in climate and the effect of nutrition. The susceptibility of insects may be greatly influenced by the quality of the food they consume. It also has been observed with some insects that a certain amount of starvation may make them more, rather than less, resistant to fumigants. In practical work it is well to know that fluctuations in resistance may occur. The alert operator must always be on the lookout for any changed conditions that may necessitate modification of recommended treatments.

Comparative Toxicity of Fumigants

Apart from the influence of the environment, there is a great variation in susceptibility of different species of insects to different fumigants. The successive stages of a given species may also vary greatly in response. The treatments recommended here are based on laboratory or field trials that have been confirmed, in many instances, by the results of successful application in practice. Note that all recommended treatments refer to specific insects or their stages or, in some cases, to clearly defined groups of insects. There is, therefore, no guarantee of the success of any attempts to apply a treatment outside the limits given in the recommended schedules.

Acquired Resistance of Insects

Many species of insect have the ability to develop resistance to certain insecticides. With fumigants this problem of resistance is a matter of increasing concern; in a global survey of stored grain pests, resistance to both of the major fumigants, phosphine and methyl bromide, was found in a number of insect species. Collections of 849 strains of insects from 82 countries showed that 20 percent of the insects had some resistance to phosphine and 5 percent to methyl bromide. The highest level of resistance (10-12 times normal) was found in the lesser grain borer Rhyzopertha dominica (F.). It was concluded from this survey that resistance to fumigants was, as yet, limited in extent and often at marginal levels, but that it was of considerable significance as it posed a real threat to the future use of fumigants as control agents.

Research in laboratories has shown that a number of destructive stored product insects can develop appreciable resistance to fumigants. Selection of the granary weevil (Sitophilus qranarius) has produced a strain with more than 12fold resistance to methyl bromide. A strain of the red flour beetle, Tribolium castaneum (Herbst), developed a 10-fold resistance to phosphine in six generations.

There is recent evidence, from field studies in India and Bangladesh, of the development of resistance to phosphine in the Khapra beetle and other stored grain pests. Resistance to fumigants is of concern because of the great value of fumigants for pest control and because of the very limited number of materials available. Even low levels of resistance in species of insects that are cosmopolitan and easily transported to other parts of the world could be of serious consequence. In view of the importance of resistance to fumigation, a brief and simplified account of some features of the problem are given below.

How Resistance Develops

When a population of insects is exposed to an insecticide some individuals are killed more easily than others. The insects that are more difficult to kill may survive after the treatment and produce offspring that are also hard to kill. These insects are said to be more tolerant because they can withstand aboveaverage doses of the poison. If a population is repeatedly treated with the same insecticide and each new generation has increasingly higher tolerance, a "resistant" strain is produced. Resistance is a genetic characteristic that is passed on from one generation to the next.

In the laboratory, resistance is produced by treating a population to kill most of the insects, breeding the tolerant survivors to produce a new generation, re-treating and continuing the process until a resistant strain is obtained. This process is known as selection for resistance. A number of strains of insects with resistance to different fumigants have been produced in this way.

In the field, resistance to fumigants can develop in the same way. In a grain bin, on a cargo ship or any other place where a resident population of insects is treated over and over again with the same fumigant, resistance might develop. Insects that are not killed may produce new generations with increasingly greater tolerance. Generally, resistance does not develop as readily in wild populations as in the laboratory because the selection process .may be irregular and because they may interbreed with nontreated susceptible insects. However, the fact that resistance has been discovered in wild populations indicates the possibility that further resistance may develop where fumigants are used regularly.

Nature of Resistance

Resistance is an inborn characteristic that allows individual insects to tolerate above average doses of poison. Resistant insects usually are similar in appearance and have the same habits as susceptible insects. Normally, they can only be distinguished by their ability to tolerate excessive concentrations of the fumigant. Tests have been designed for detecting and measuring resistance to fumigants.

Testing for resistance

For routine monitoring to detect the initial appearance of resistance in wild populations of stored product beetles, it is convenient to use a discriminating dose, which is expected to kill all susceptible specimens. The dose chosen is that corresponding to slightly above the LD(99) 9 obtained from the regression line for susceptible beetles allowing for, in the case of phosphine, what appears to be inherent variability of response.

When using a discriminating test with fumigants it is always advisable to make provision for abnormal concentrations. If a concentration is obtained that is less than the discriminating concentration, this will be revealed by abnormal survival in the susceptible reference strain. Abnormally high concentrations may be revealed by the inclusion in the tests of a reference strain (or species) with slightly greater tolerance to the fumigant than the susceptible reference strain on which the discriminating dose is based, approximately x 1.5 for methyl bromide tests and x 2.5 for phosphine tests. An alternative approach is to use three dosages, one at the discriminating dose, one at the approximate LD(90) level and the other at an equivalent level above the discriminating dose.

In regular monitoring for resistance, it should be detectable even when only a small proportion of resistant individuals is present. A minimum of 100 insects in two batches of 50 should be used per sample. Limited numbers of insects may not be sufficient to detect low levels of resistance. Therefore, additional samples should be obtained, if possible. If, however, there is suspicion of serious resistance (e.g. from failure of treatments) a test with small numbers (10 to 20) may provide valuable early indication. The insects are exposed to the discriminating dose for the appropriate period in the usual way. If all of them are dead at the end of the posttreatment holding period, the sample can be classified as "no resistance detectable", and the medium in which they were held is put into a hot-air oven to destroy the culture.

Confirming resistance

The appearance of unaffected insects in a discriminating test could be due to the presence of unusually tolerant individuals from a normal population. Provided that the conditions of exposure, the physiological state of the insects and the dosages are consistent, the probability of a single insect in a batch of 100 being unaffected due to chance is less than 0.1 (e.g. less than once in 10 tests). It is important to determine whether incomplete response is due to such causes or to genuine resistance. This can be checked as follows:

1. The test can be repeated using further samples from the same field population. The chances of adventitious failure to respond by a single individual in each of successive tests decline progressively (less than 0.01, 0.001, 0.0001 and so on). Survival of two or more indviduals throughout is even less likely. Therefore, the continued appearance of a proportion of unaffected individuals can be considered as proof of resistance. 2. Alternatively, the insects which were unaffected in the discriminating test may be kept and used for breeding a further generation. If their reaction is actually due to resistance, it will be found that a substantially larger proportion of their progeny will fail to respond to the discriminating dose.When these tests indicate that a population of insects is resistant, then extensive testing should be carried out to determine the degree of resistance present.

Way to avoid resistance

Precautions can be taken to reduce the possibility of insects developing resistance to fumigants.Perhaps the most effective measure involves alternate control practices that do not require chemicals. Good sanitation procedures, proper storage conditions, insect resistant packaging and all other measures that prevent infestations from developing can do much to reduce the need for fumigants. Treatments such as aeration of the commodity, irradiation, temperature extremes, insect pathogens, etc. Where fumigants have to be used on a regular basis, close guard should be kept against control failures.

Complete control of all insects in a treatment is the best insurance against resistance. Periodic checks for resistance should be made in areas that are fumigated regularly. If signs of resistance begin to appear (as indicated either by control failures or through the test procedure) then every effort should be made to eradicate the population. The measures necessary for eradication will vary in different situations; they may involve a number of procedures using both chemical and non-chemical means. Rotation of fumigants may be effective in some instances, especially if crossresistance is not a problem. For example, methyl bromide might be used at intervals in a control programme that relies mainly on phosphine.One measure that is not advisable in dealing with resistance problems involves increased dosing. Such practices as doubling the dose of fumigant to achieve an economic level of control can magnify the problem unless complete eradication is assured. Any insects surviving increased doses may develop even higher levels of resistance than would occur with the normally recommended treatment.

FIELD DETERMINATION OF FUMIGANTS

A number of instruments are available on the market for analysis of fumigants under practical operating conditions. Determinations may be conducted at regular intervals both in the free space and in the commodity. Used in conjunction with integrated concentration—time products, such analyses enable the operator to monitor concentrations throughout a treatment and know when the desired measure of treatment has been attained in all parts of the system. The fumigation may then be terminated at the appropriate time. Apart from the determination of full fumigant concentrations during actual exposure, much of the equipment may also be used to measure the success of the aeration process as indicated by the presence or absence of residual vapours. Some of the equipment may also be used for the purpose of detecting leaks from the structure during treatment.

Thermal Conductivity Analysers

In recent years a portable instrument known as the thermal conductivity analyser or meter has been used extensively for fumigant determination, principally with methyl bromide. This was first developed for practical use by Phillips and Bulger.

Principle of operation

The basic principle underlying this instrument is that when a constant electric current is passed through a wire, the final equilibrium temperature of the wire is affected by the composition of the gas surrounding it. If the composition of the gas is changed, the equilibrium temperature of the wire will alter. This in turn will alter the resistance of the wire. In a thermal conductivity apparatus for gas analysis, a Wheatstone bridge circuit is used to measure the imbalance caused by passing gas over the detector filaments. There are usually four or eight filaments in the same number of cells. Half of the cells are used for passing the fumigant/air mixture and the other half, in which only air is present, are used as a control. When an electric current is passed through the filaments, the whole bridge is balanced if the composition of the gases surrounding all the filaments is the same throughout. If the cells surrounding the detecting filaments are filled with a different gas mixture, the bridge becomes unbalanced; the extent of this can be measured by a galvanometer. By calibration with known concentrations of a given gas the galvanometer readings can be transposed into the units of concentration desired, such as 9 per m^3.

Standard equipment

A commercial thermal conductivity analyses has the following components:

1. Four tungsten filaments in as many cells, a pair providing each arm of the bridge. The cells are bored in a brass block. Two cells are used as a control to hold the standard gas, which is air, and the other two are incorporated in the sampling train of the gas-air mixture undergoing analysis.
2. A galvanometer from which readings are made.
3. A potentiometer for current control across the filaments.
4. A separate gas passage for drawing samples through the cells, with inlet and outlet connexions.
5. A source of electric current, which may be provided by batteries contained in the instrument or by connexions to outlets from the local main supply. Direct current of 6 volts is used in the instrument and, if the main supply is used, transformers and rectifiers are needed.

In addition some instruments may contain one or more of the following pieces:

1. An aspirator with a rubber hand bulb for drawing a constant flow of the sample across the cells.
2. An electrically driven pump for the same purpose as in (1).
3. A flowmeter for use when the mechanical pump is used.
4. A guard tube to hold soda-asbestos, or similar material, used to remove water vapour and carbon dioxide from the incoming samples.

In some instruments the reference cells are permanently closed, whilst in others they are open. Sometimes the reference cell is protected by a guard tube, but this is sealed off when sampling begins. Ideally, it is desirable to take only small samples at a time and an instrument operated by a hand bulb ensures this. When a large structure is undergoing treatment, however, samples have to be drawn from considerable distances and mechanical pumps are necessary. In practice, the thermal conductivity analyses is unsuitable for use with mixtures of fumigants. It may be possible to calibrate the instrument to indicate concentrations of a mixture in a flask or chamber containing the gaseous mixture only, but in the presence of material being fumigated, the various components of the mixture would be sorbed at different rates and the readings would not provide an accurate indication of the relative proportion of each fumigant present in the free air.

Tupes of thermal conductivity instrument

There is a range of instruments available on the market, which vary in accuracy and cost according to the quality of the components incorporated. A bulletin by Heseltine described in detail the construction and operation of a battery-operated meter, now obtainable commercially, which is of sufficient accuracy to be used in the laboratory and the field.

Calibration

It is most important that the thermal conductivity analyser be calibrated frequently against a known concentration of the fumigant or fumigants for which it is being used. Kenaga described a simple apparatus, using carbon tetrachloride as the standard gas, for the calibration of the thermal conductivity instrument for various fumigants. Carbon tetrachloride gives the same galvanometer reading as methyl bromide, and since it can initially be measured as a liquid at ordinary temperatures, it is more suitable for calibration. It is also advisable to check periodically the performance of a thermal conductivity analyses under field operating conditions by taking a series of samples for chemical analysis and comparing the results with instrument readings corresponding in position and time to the origin of the samples. The instrument is then adjusted according to the results of the chemical analysis.

Interference refractometers

Instruments designed to utilise differences in the refractive index of gases have been employed successfully for determining fumigant concentrations. In this type of equipment, parallel light from 8 collimator is divided into two beams by two slits and passed through two tubes, into one of which has been introduced the gas mixture under test. The tubes are closed by opticallyworked glass plates. On emerging from the tubes the two separate beams are brought together by a lens and thus produce in the focal plane of the lens very fine vertical fringes, which can be viewed through an eyepiece. After the zero reading has been set in both tubes in ordinary air, a sample of the atmosphere containing the fumigant under test is drawn into one of the tubes by the squeeze bulb and the difference in the refraction of the gases in the two tubes, as shown by a shift in the fringes, is measured on the scale. By suitable calibration of the readings for a particular fumigant gas, the percentage concentration in the atmosphere under analysis may be easily measured. For greater accuracy in making readings, some operators have found that insertion of a piece of glass capillary tube in the bulb tube will regulate the inflow of gas so that the chosen fringe does not move off the scale. Since the brightness of a fringe can vary according to its position on the scale, this ensures that the same selected fringe is used at all times.

Theoretically, an instrument employing this principle gives an absolute reading and is not subject to variable conditions, such as variations in voltage or the failure of component parts to function accurately, which may be encountered with other types of instruments. In practice, such an instrument is simple to operate and readings are reproducible under uniform conditions. However, in common with all apparatus used under field conditions, initial accurate calibration is essential. Instruments that give different concentration ranges and different degrees of sensitivity are available, the price increasing with the sensitivity of the equipment.

Detector tubes

Glass detector tubes used for determining the concentrations of a wide range of gases in air are available on the market. These tubes are particularly suitable for use with fumigants which may present a fire hazard under conditions in which a device such as the halide lamp would present a hazard. The use and accuracy of two makes of these tubes for a number of different gases have been discussed in detail by Dumas and Monro.

Colour indicators

Colour indicators have been developed commercially for fumigant determination, more particularly with the use of ethylene oxide as a sterilising agent. These indicators are tapes placed in or on the material being sterilised, or they may be small sachets containing chemicals which react proportionately to the intensity or duration of exposure. An automatic toxic gas detector that utilises indicator tapes has been developed for HCN and other toxic gases. Heseltine and Royce described the application of sachets for both ethylene oxide and methyl bromide fumigations. According to these authors, the sachets used for ethylene oxide may be inspected for the appropriate colour change, either through a window in the treatment chamber or by withdrawal. The methyl bromide sachets give no direct colour reaction and it is necessary to carry out a titration following withdrawal from the fumigation system. With both gases the sachets may be used to determine if a desired concentration x time product has been reached in any part of the system thus ensuring that control of the insects or other organisms has been achieved.

Lamps

Detector lamps are used, at present, exclusively for halogenated hydrocarbon fumigants. Their best use is for detecting leaks from the system and as a safety check during aeration.

Halide meters

Instruments for measuring low concentrations of halogenated vapours in air, utilising the principle of photometry, are commercially available. The intensity of the blue lines of the copper spectrum, produced in an electric arc between two electrodes, is continuously measured with a photo-electric photometer using a blue-sensitive phototube fitted with a blue glass colour filter. Halide vapour coming in contact with the hot tip of the copper electrode reacts to form a copper halide, which vaporises at the temperature of the electrode and is carried into the arc. The intensity of the blue spectrum is proportional to the concentration of halide vapour present.These instruments are primarily designed for measuring halogenated hydrocarbons in air from 0 to 500 parts per millions (ppm) with 10 percent accuracy. They are used mainly for safety purposes, but Roth has found an instrument of this type useful for measuring concentrations of ethylene dibromide up to 7 mg/l in commercial fumigations. For higher concentrations a simple dilutionsampling technique is necessary.

Infra-red (IR) analysers

These instruments can be used for analysing the high concentrations of fumigants needed for insect control as well as the lower levels that may contaminate atmospheres in the work place. They are portable, battery powered, direct reading and have no flame; they can be used safely in dusty atmospheres and are useful for determining whether a space is safe for occupancy.

Gas chromatographs

Portable gas chromatographs are available for gas analysis in field operations. This instrument is suitable for the high concentrations used for insect control as well as for low concentrations around the threshold limit value for human health. Although these instruments are expensive, they are accurate and relatively easy to use under field conditions.

FUMIGANT RESIDUES

When a pesticide residue remains in food or food products, several factors will determine its importance as a hazard to human health. The average fraction of the total diet likely to contain food with this residue is important, as well as the nature and toxicity of the residue itself.

The following definitions of the terms used in work on pesticide residues are given by the joint FAD/WHO Committee on Pesticide Residues:

- *Residue:* a pesticide chemical, its derivatives and adjuvants in or on a plant or animal. Residues are expressed as parts per million (ppm) based on fresh weight of the sample.
- *Food factor:* the average fraction of the total diet made up by the food or class of foods under discussion. Details of the diet of a country may be obtained from the FAO food balance sheets or other similar data. Acceptable daily intake: the daily

dosage of a chemical which, during an entire lifetime, appears to be without appreciable risk on the basis of all the facts known at the time. "Without appreciable risk" is taken to mean the practical certainty that injury will not result even after a lifetime of exposure. The acceptable daily intake is expressed in milligrammes of the chemical, as it appears in the food, per kilogramme of body weight.

— *Permissible level:* the permissible concentration of a residue in or on a food when first offered for consumption, calculated from the acceptable daily intake, the food factor and the average weight of the consumer. The permissible level is expressed in ppm of the fresh weight of the food.

— *Tolerance:* the permitted concentration of a residue in or on a food, derived by taking into account both the range of residue actually remaining when the food is first offered for consumption (following good agricultural practice) and the permissible level. The tolerance is also expressed in ppm. It is never greater than the permissible level for the food in question and is usually smaller.

Nature of Fumigant Residues

The kind of residue left after a fumigation may consist of original fumigant, reaction products formed by a combination of fumigant with components of the commodity or end products of a formulation that generates the fumigant. Unreacted fumigant can remain in some materials for appreciable periods after the treatment. Usually, the amount remaining decreases progressively with time; however, some highly sorptive fumigants such as carbon tetrachloride, ethylene dibromide and hydrogen cyanide may persist in some materials for weeks or months after airing.

Some fumigants react with components of commodities to form new compounds. Ethylene oxide can combine with the chlorides and bromides in food to form toxic chlorohydrins and bromohydrins. Methyl bromide is decomposed in wheat to form several non-toxic derivatives and hydrogen cyanide can combine with sugars in dried fruit to form laevulose cyanohydrin. Other fumigants may also react with materials being fumigated. In addition to residue from the fumigant, some by-products from formulations such as aluminium phosphide and calcium cyanide can leave residue on food materials. An ash-like residue of aluminium hydroxide, along with a small amount of undecomposed aluminium phosphide, is left after phosphine is generated. Calcium cyanide leaves a residue of calcium hydroxide after hydrogen cyanide is released.

Significance of Fumigant Residues

The residues remaining in treated materials after a fumigation may be of significance both as an occupational hazard to workers and others exposed to desorbing gas and as a hazard to consumers eating treated foods. Although desorbing fumigant may not be considered a residue in the usual sense, appreciable amounts can remain for long periods of time and create hazards for personnel in the immediate vicinity. When treated goods are kept in confined spaces, such as airtight bins or a ship's hold, the residual fumigant can be of considerable consequence. There is great concern over the possibility of longterm effects that may develop from exposure to desorbing fumigant.

Some fumigant may remain in food materials and reach the ultimate consumer. Attention has been focused on residues of pesticides in food in recent years because of the harmful effects they may have on human beings. Concern over toxic chemicals in food has been heightened by sensitive detection methods that show traces of residue not previously suspected. The significance of very low levels of some compounds is not known. However, it is believed that the human body can tolerate small amounts without adverse effects. Therefore, residue tolerances are established on the basis of extensive investigation of toxic hazards. It should be pointed out that numerous surveys for fumigant residues on food have shown only low levels in just a few samples and that cooking normally reduces these to even lower levels. However, concern has been expressed for fumigated foods that are not normally cooked and special recommendations have been given for these situations. In addition, residues that affect food quality through offensive odours or other factors may be of significance. Good fumigation practice will normally require that treatments should be conducted in such a way as to keep residues to the lowest possible level.

Factors affecting residue accumulation

The amount of residue that remains in fumigated materials is determined by the conditions existing during the fumigation and the treatment of the material afterwards. In some cases residue levels may be held to a minimum if the various factors that lead to residue accumulation are taken into account before the treatment is done. A few general statements on residue accumulation can be made. However, it must be emphasized that no one condition is likely to apply equally for all fumigants or for different commodities.

Type of Fumigant

Fumigants with high boiling points tend to be sorbed to a greater extent and remain as residues longer than more volatile compounds. For example, acrylonitrile was found to remain in wheat for many days, whereas methyl bromide dissipated in a few hours. Fumigants that react with plant or animal constituents may also leave appreciable residue. This may be fixed residue such as inorganic bromide, chloride, phosphate or other compound, depending on the fumigant, or it may be a volatile material such as ethylene chlorohydrin from ethylene oxide or dimethyl sulphide from methyl bromide.

Solubility in water can also influence residue accumulation. HCN is not used on some moist materials because of the burning effect of the acid formed when it combines with water.

Type of commodity

Some materials will sorb and retain more fumigant than others. Foods with high oil and fat content may retain more residue than cereals. Rhodes et al indicated that methyl bromide is readily absorbed by lipid materials and they suggest that care should be taken to avoid contamination of high fat content foods such as butter, cheese, margarine, meat etc.Different fractions of seeds contain different amounts of residue. The shells of walnuts have been found to contain 70 percent of the total residual bromide remaining after fumigation with methyl bromide,A substantial portion of residual carbon tetrachloride in treated wheat appeared in milled fractions, especially the bran. The gluten fraction of wheat flour contained 80 percent of the total decomposed methyl bromide. Finely divided materials can often absorb more fumigant and retain more residue than whole seeds. Some materials are not treated with certain fumigants because of the reaction products that remain as residue. Thus, sulphur-containing goods are not treated with methyl bromide, and materials containing copper or copper salts may react with phosphine, depleting concentrations of the fumigant from the atmosphere and forming undesirable residues.

Concentration and exposure time

The amount of residue accumulating during a fumigation can be influenced by the dosage applied and the length of time the material is exposed. The amount of bromide retained by citrus fruit after fumigation with methyl bromide and ethylene chlorobromide WAS found to be greater with higher dosages and longer exposure times. Similar observations have been made on other commodities treated with methyl bromide, carbon tetrachloride, ethylene dichloride and ethylene dibromide.

Moisture content and humidity

The retention of sorbed gases and the reaction of fumigants with components of treated goods are influenced by the moisture content of the goods and by the relative humidity of the sir around them. Usually, sorption is higher in materials with higher moisture content. In dried fruit fumigated with HCN, moisture content was found to be the main factor governing retention of cyanide; fruit of 19 percent moisture content retained four times as much free cyanide and had eight times as much laevulose cyanohydrin as fruit of 8 percent moisture content. Maize at 15 percent moisture content retained twice as much ethylene dibromide as at 9 percent. Humidity of the atmosphere also appears to be an important factor in the dissipation of fumigant. Greater desorption of ethylene dibromide from layers of wheat was found to occur at high rather than low humidities.

Temperature

The rate of desorption of fumigant is usually related to temperature, with less abreacted fumigant residue remaining at high temperatures. However, residue from chemical reaction is likely to be greater at higher temperatures. The inorganic bromide residues in flour increase with increases in temperature, even when the dosage is decreased.

Multiple treatments

If commodities are refumigated with some fumigants, the level of residue may be expected to increase with each treatment. Cereal grains given repeated treatments with methyl bromide were found to contain increasingly higher levels of inorganic bromide both in the whole grains and in the flour milled from them. Similarly, flour fumigated several times with methyl bromide has more residue after each treatment. On the other hand, flour refumigated with phosphine contained no more measurable residue than when only treated once. Since there is considerable possibility of re-infestation and subsequent retreatment of goods in international trade, and the history of such treatment may not be known, some precautions may be needed to ensure that the residue levels do not exceed permitted tolerances.

REFERENCES

Dumas, T. "Inorganic and organic bromide residues in foodstuffs fumigated 1973 with methyl bromide and ethylene dibromide at low temperatures" .*J. Aqric. Food Chem.*, 21 : 433 - 436.

Gray, H.E. Vikane, "A new fumigant for control of drywood termites". *Pest 1960 Control,* 10 : 43 - 46.

Kenaga, E.E. "Evaluation of grain fumigants". *Down to Earth,* 13 : 10 - 13. 1957a

Leesch, J.G. & Highland, H.A. "Fumigation of shrink-wrapped pallets". 1978 *J. Georgia Entomol. Soc.*, 13 : 43 - 50.

13

Socioeconomic Aspects of Soil Degradation

Soil degradation is a major global environmental problem, causing widespread and serious impacts on water quality, biodiversity and the emission of climate changing greenhouse gases. The chemical and physical deterioration of soils also has major implications for agricultural productivity. According to a recent study, nearly 40 percent of the world's agricultural land experiences serious productivity impacts due to soil degradation, with rates up to 75 percent for some regions. Yet a primary cause of soil degradation is the depletion of soils which results from the farming systems chosen by the farmers. Decisions such as where and when to produce, the types of techniques used—particularly in land preparation—and the level and timing of inputs, all affect the biophysical quality of the soil, and the extent to which it is depleted, conserved or augmented. Thus, although society and farmers themselves may benefit from higher levels of soil quality, it has proven difficult to achieve the changes in farm management practices necessary to reach this goal.

Although soil degradation is generated by farmers of all income levels, taking a closer look at poor producers in particular is warranted because of the impact of poverty on farmers' production decisions, and because of the potentially important role of improved soil management in poverty alleviation. Agricultural production is the most important source of income to a majority of the world's rural poor, who currently number around 800 million. Through its effect on agricultural productivity, the quality of their soil resources has a major impact on the capacity of poor farmers to achieve food security. Improvement of soil resources thus in many cases represents an important avenue for improving incomes among the world's poorest inhabitants.

NATURAL RESOURCES IN AGRICULTURAL PRODUCTION

Capital is a stock of real goods with the potential to produce a flow of benefits or utilities in the future. Natural capital, then, is the stock of goods derived directly from nature that have the potential to contribute to the long-term economic productivity and welfare of societies. It includes raw materials such as timber, water and soil, as well as environmental services such as waste assimilation and watershed maintenance. In addition, natural capital provides utilities through the provision of aesthetic and recreational services.

As with financial capital, natural capital is measured in stocks and flows, although in physical rather than monetary units. Natural capital stock and flow values may be converted to monetary units with the application of resource prices to the physical quantities, although this exercise is often problematic due to imperfections in resource markets that lead to distorted prices. Adjustments to market prices to reflect the true opportunity costs of resource use to society are therefore necessary, but difficult to assess and often controversial.

Natural capital stocks are commonly divided into renewable and exhaustible categories. The former is defined as "a plant or animal population with the capacity for reproduction and growth at a significant rate when viewed from man's economic time scale". These include fish populations, forests and, under certain circumstances, soil fertility. Such resources are capable of regenerating themselves—as long as the environment in which they are nurtured is favourable. Upsets in this nurturing environment may lead to a loss of regeneration capacity and thus the exhaustion of the resource. Exhaustible resources on the other hand, are those which are limited in quantity and not producible in any economically relevant time frame. Extraction of one unit of this resource results in a decrease in the total stock of the resource by exactly the same amount. Natural resources that fall into this category include hydrocarbons, groundwater from aquifers and minerals.

The distinction is important, as it drives the "optimal" rate of investment, when viewed from the neo-classical economic point of view. In the case of exhaustible resources the key factors determining the optimal depletion path are the cost of "harvesting" the resource, its price and the discount rate. For renewable resources, the natural rate of regeneration enters into the calculation. Human intervention can change the natural rate of regeneration through capital investments augmenting the stocks, as well as through depletion, which adds another dimension of complexity in determining economically optimal patterns of use.

Economists have traditionally approached the analysis of natural resource use and allocation over time as an investment decision. However, in the application of capital investment analysis to systems that include natural capital it is necessary to recognise that there are fundamental ways in which natural capital differs from financial capital. The first is that natural capital provides services to humans (and other life forms) other than providing raw materials in the production process. Natural capital provides waste assimilation services as well as other ecosystem functions which provide utilities such as recreation, health, cultural and aesthetic services, as well as the maintenance of essential climatic and ecological cycles and functions. These additional functions of natural capital must be considered when assessing the use and savings of natural capital.

Another unique characteristic of natural capital is that it is produced and maintained within a complex web of biological and physical relationships that are defined as ecosystems. Ecosystems are "communities of organisms in which internal interactions among the organisms determine behaviour more than do external biological events". Ecosystems occur at various scales—from microscopic to global, depending on the type of behaviour that is being analysed. Often there is a high degree of linkage between the functions of various species within an ecosystem, so that disturbances to one member of the community lead to impacts throughout the system, which in turn may impact the ability of the system to maintain its functions.

Ecosystems are characterised by their degree of biological diversity and resilience. The two are linked; higher levels of biodiversity generally give rise to a higher degree of resilience. Resilience is defined as "the ability of the ecosystem to maintain its characteristic patterns and rates of process in response to the variability inherent in its climate regimes. The impacts of depleting one or more natural capital assets on the various ecosystems in which it is embodied vary tremendously across resource and ecosystem types.

By its very nature, the analysis of natural resource investment and management requires an intertemporal dimension, as resource flows are dynamic. Changes in the stocks of renewable and non-renewable resources in the current time period have implications for the future flows of the resource. Thus the use of natural resources must be considered as a trade-off over time periods. For non-renewable resources, current consumption of the resource may result in higher marginal costs of extraction for future users, e.g. depletion of a groundwater aquifer resulting in higher pumping costs to future users of the well. With non-renewable resources, tracing the changes in the stock of the resource over time is often a simple matter of accounting for the amounts withdrawn or added to initial stocks. However, with renewable resources, the natural regenerative process of the resource and its relationship with the initial stocks must be considered, as well as the harvest rates over time. In some cases this may be a quantity change over time (e.g. fish population levels) while in others it may be a quality change over time (e.g. soil fertility).

An additional feature of natural capital that is critical to consider in an investment analysis is the fact that investments are partially or entirely irreversible. This is true of most capital investments: sunk costs are generally irretrievable. However, in the case of investment strategies involving natural capital, a larger dimension is involved, due to the embeddedness of natural capital within an ecosystem. Changes in the level of natural capital stocks can result in changes in entire ecosystems and thus their ability to provide ecological services. This is clearly the case with the depletion of an exhaustible resource, but may also occur with renewable resources, when the patterns of use exceed the capacity of the system to renew itself. In such cases the resilience of the ecosystem is driven below a threshold resulting in a major transformation of the system—and the loss of some of its characteristics and functions.

Incentives for Investment

In the course of managing agricultural production and the natural resources involved therein, farmers constantly make decisions about the relative merits of various investment and depletion options regarding their natural capital assets. These decisions are made under uncertainty: the farmer does not know what future market and production conditions will be, and in many cases does not know how his/her actions may be manifested in environmental outcomes. According to orthodox economic theory, if farmers are "rational", e.g. motivated by profit, their decisions will be based on an implicit or explicit calculation and comparison of the expected stream of benefits and costs associated with the activity over time. An important determinant of the decision will be the rate at which the decision-maker trades off current for future consumption, or the discount rate. If the discounted difference between costs and benefits, e.g. the net present value, of the investment being considered is positive, then it should be undertaken.

Even in this relatively simple framework for analysing investment incentives there are several complicated issues to be resolved. One key issue is how people's discount rates are determined, and the implications of the rate for investment decisions related to natural capital use. Another major issue is how the presence of uncertainty and risk affects the investment decision. These issues are discussed below.

Discounting is the most common way in which economists account for the time dimension of investments. The discount rate is an indicator of the trade-off between present and future consumption. Often, the interest rate on capital is used as an inverse indicator of the discount rate, as it is a market-determined price for the exchange of goods or services across time. The choice of discount rate has an important impact on the analysis of incentives for investment, because the costs and benefits associated with management are incurred at various points in time and the discount rate determines how this temporal distribution has an impact on the desirability of the investment. For example, a high discount rate indicates a preference for present over future consumption, and thus disincentives to adopt management strategies that result in delayed returns.

The implications of how discount rates have an impact on incentives for sustainable natural resource use are quite controversial. Some argue that because of the bias against investments with delayed returns, high discount rates universally work against sustainable natural resource management, and thus decision-makers with higher discount rates are more likely to be engaged in environmentally degrading activities.

Norgaard and Howarth provide an interesting perspective on the relationship between discount rates and sustainable resource management. First of all, they point out that a high discount rate could result in environmentally beneficial outcomes as they make capital investment projects which may have negative environmental impacts—such as a dams—less attractive. Secondly, they argue that the transfer of resources to future generations should rightly be considered as an equity issue, and not one of efficiency. The discount rate is a way of determining efficient resource allocation over time, given the fact that goods and services have a time value associated with them.

Given a time value of money that investors are faced with, e.g. interest rate on capital, it is appropriate to discount when looking at the efficiency of investments from the current generation's point of view. This decision is fundamentally different from a decision on resource allocations which is based upon equity considerations. In this latter case, a decision-maker may choose to bequeath resources to future generations in order to achieve a desired level of income or natural resource distribution. Norgaard and Howarth distinguish between these two types of resource allocation decisions, but make an important link between them: resource allocations made between current and future generations based upon equity considerations will change the efficiency prices which current generation investors face. The desire to transfer assets to future generations based on equity considerations has an impact on the allocation of resources over time and, thus, the investors' discount rate.

Estimating the discount rate among decision-makers is often problematic. It is quite tricky to estimate the rate at which individuals value future versus current consumption, and market-derived indicators, such as the interest rate, are often subject to distortions, limiting their ability to reflect true preferences. Individual investor's discount rates may change over the course of their life: younger people may be more willing to wait for future consumption than the elderly whose time horizon is shorter. Discount rates may vary across different types of investments, if the fungibility of assets held by the household is constrained. Another issue is that the rate at which individuals trade off present and future consumption is usually assumed to be different from that of groups or society, so the discount rate applied under a joint investment scheme may be quite different from that of a private investor.

Investment under Uncertainty

The presence of risk has a significant impact on the choice of investment and natural resource management strategies. It is necessary to consider both the nature of the risks encountered as well as the attitude of the decision-makers towards risk and their capacity to insure against risks, in order to capture the impacts on investment behaviour.

Risk arises from the presence of uncertainty and can be categorised by several different features: their frequency, the intensity of the impact they create, the degree to which they are experienced on an individual versus a collective basis, and the degree to which they are correlated over time. These characteristics are an important determinant of the ability of decision-makers to cope with risk, either through ex-ante insurance schemes, or ex-post consumption smoothing mechanisms. Either of these may have significant impacts on investment decision-making.

Producers' attitudes towards risk, and the degree to which they are averse to being exposed to risk have an impact on the types of investments they are willing to undertake. Results from theoretical and empirical studies indicate that producers who are more risk averse will allocate resources to activities that are less risky (e.g. lower variability of returns) but which have a lower average return.

Risk aversion not only affects the composition of the portfolio of activities undertaken, but also the production levels within activities. Theoretically, risk averse producers may be expected to produce less when exposed to risk. However, the impacts of risk on investment decision-making can be quite different if the consumption and production decisions of the decision-maker are linked. Many small-scale agricultural producers are both consumers and producers of agricultural production, making the consumption and production decisions non-separable. Sadoulet and de Janvry note that risk-averse producers may actually increase production of a risky commodity if they are net buyers of the commodity. Grepperud points out that it is necessary to know how inputs affect production risk, and distinguish production from market uncertainties, in order to determine how risk will impact farmers' soil management choices.

Risk preferences vary among populations also due to the fact that people have different perceptions of the probability of an event occurring, which may be due to differing degrees of information available to each. These differences in risk perceptions and preferences give rise to the possibility of trading between individuals for insurance services or for commodities with differing degrees of risk associated with them. However, in order for such markets to be efficient in coordinating decisions involving time and uncertainty, they would have to include all commodities at every date, and in every possible state, or *contingent commodities*. Under such conditions the presence of risk would not affect investment decisions, as managers could efficiently insure against it. Since a complete set of contingent commodity markets clearly does not exist, and the ability of individuals to insure against risk is not perfect, risk preferences do enter into the investment decision.

In addition to the issues raised in a standard investment analysis framework, the impact of irreversibility on the incentives to make investments is an important aspect that needs to be considered in the context of natural resource management. Frequently, indeed usually, the potential future value of holding a natural resource stock is uncertain, especially if all of the services it does and may provide are included into the calculation. The presence of this uncertainty results in a cost associated with irreversible investments or depletions (e.g. negative investments) The combination of irreversibility of investment with the uncertainty over the value of future losses results in an "option value" of waiting for more complete information before any irreversible action is taken.

There are two important aspects of irreversibility that may have an impact on a farmer's investment decision under conditions of uncertainty. In the first case, irreversibility arises from a lack of investment or depletion of the resources, which results in irreversible loss, which can be true for renewable resources as well as exhaustible ones. In this situation, there is a value associated with investments that result in the avoidance of an irreversible loss (e.g. extinction), with the result that investments are more attractive to the decision-maker. In the second type of example, the investment represents a "sunk cost" which is irreversible from the farmer's point of view—he or she cannot recover the capital that went into such an investment. In this situation, there is a positive value to waiting to accumulate more information before making such an investment, particularly in the case where liquidation of productive assets is a form of insurance employed by the producer.

The impact of irreversibility thus has two opposing effects on farmer's investment incentives: on the one hand a positive incentive to avoid irreversible loss of natural capital services, and on the other a negative incentive for an immediate commitment of scarce investment capital that cannot be recovered. The relative weight of these two effects will depend on the extent to which an irretrievable loss of natural capital services is experienced as a private cost to the farmer, rather than an externality, and the degree to which the farmer prefers to keep liquid assets versus sunk capital assets.

Externalities and Investment Incentives

Many of the impacts generated by the investment or disinvestments of farmers in natural capital are realised in the form of externalities. When an agricultural producer takes an action that results in costs (benefits) that are borne by others, an externality is generated. Examples include the siltation of downstream waterways with eroded soils from upstream producers, the pollution of water sources from agricultural chemical runoff, the loss of biodiversity associated with land clearing for agriculture and the generation of greenhouse gases which contribute to global climate change arising from the loss of biomass associated with some types of agricultural production. Externalities can occur over both space and time, e.g. between current and future generations or between individuals and groups distributed over space.

External costs and benefits arising from agricultural production are distinguished from private ones—whose impact is felt by the agent taking the management decision. The distinction is quite important, as it implies that the full value of the action is not considered in the decision-making process, thereby resulting in a suboptimal amount of resource depletion or investment. Externalities are an important form of market failure, and their existence implies that a strict reliance on market forces will not yield an efficient allocation of resources. For this reason, some sort of intervention is required in order to correct the market failure and in order to reach a more efficient allocation of resources. Correcting the market failures associated with externalities will result in an increase in social welfare, as resource investment and depletion decisions will be taken in accordance with the true social values of the resources in question.

One of the primary reasons for the generation of negative environmental externalities through agricultural production (and other sectors) is ill-defined property rights to the resources or environmental services. If property rights were well defined and enforced, the owners would have the incentive to make investments whose future values can be expected to accrue to them. In addition, the presence of a well-defined property rights structure creates the possibility of trading among owners, e.g. a market for the goods, and thus an efficient allocation of resources could be obtained. There are several reasons for the lack of a well-defined property regime with regard to natural resources and environmental services. One main issue is that it is often difficult to quantify the resource, due to a lack of technical understanding of the good, as well as heterogeneity in the distribution of the resource over space and time. In many cases, natural resources require collective action among claimants under common property or private property regimes, which is difficult to achieve, due to high transactions costs and socio-economic pressures such as population growth and land use changes which create conflict rather than cooperation. In some cases, the environmental good or service in question has a public good nature, which means that exclusion of access is not possible, thereby reducing the incentives of any one claimant to invest in its provision.

From the perspective of analysing investment behaviour, the critical feature of all types of externalities generated through agricultural production choices is that they do not enter the investment decision-making process of the producer. However, it is also important to recognise that, in many cases, the same actions which generate negative environmental externalities also result in some impacts on the producers themselves, and to this extent the costs will be considered in investment and production decision-making. Assuming the producer is a profit maximiser, decisions on investing or depleting in natural capital will then be guided by their impact on the total profit making capacity of the producer. It is therefore important to understand the relationship between natural capital and agricultural production outcomes, in order to understand the rationale of producers' investment and production decisions.

It is also important to understand that when market failures are present, the production and investment choices made by farmers will not lead to a socially optimal allocation of natural capital and, in order to reach this goal, it will be necessary for some sort of intervention. In recent years considerable interest has been raised in programmes and policies which will provide incentives to agricultural producers to adopt practices which will generate local or global environmental benefits. One such example is the Global Environment Facility which supports projects that provide incentives to agricultural producers to adopt land use practices that conserve biodiversity, protect watersheds and reduce greenhouse gas emissions. The Clean Development Mechanism which is being negotiated under the Kyoto Protocol is a potential example of a market based mechanism for transfer payments to land users for the adoption of practices which generate carbon sequestration benefits. On a more local scale, power-generating facilities may pay upstream farmers to adopt practices that reduce erosion, in order to obtain higher hydropower generation capacity.

These transfer mechanisms have a tremendous capacity for generating both on-farm and external production and environmental benefits. However, it is critical to have a clear understanding of the farm level perspective on the benefits provided by natural capital, as well as the costs of investment and the various types of market and non-market barriers that farmers face when contemplating investments in natural resources in order to design effective strategies for improving the welfare of farmers as well as others in local, global or future communities.

NATURAL CAPITAL IN AGRICULTURAL PRODUCTION

For any agricultural producer, natural resources play a key role in the production process. Agricultural production is a process of transforming natural resources into a form that is usable to humankind through the application of labour and capital inputs. The stock of natural capital available to the farmer at any one point in time will have an impact on the rate at which this transformation occurs, e.g. the productivity of the system. Conversely, the nature of the agricultural production practices employed—including the investment decisions made by the producer, has an impact on the stock of natural capital and thus future flows of natural capital and agricultural production outcomes reliant upon such flows.

Natural capital can thus be thought of as an input to agricultural production, alongside the more commonly thought of inputs such as labour, financial capital and technology. Elements of natural capital which are most relevant in the context of agricultural production include soil quality (slope, texture, composition) climate (amount and timing of rainfall, solar radiation, wind), water resources (availability, quality) and land cover (amount and type of vegetation).

Agricultural production levels are one of the major determinants of rural incomes. Production variance determines the degree of production risk producers face, which is also a critical determinant of consumption levels, as well as investment strategies. Input efficiency is an important determinant of production costs and thus net returns to production. While there are many other benefits which farmers may consider in making their soil management decisions, these three aspects are the major determinants of the financial and risk benefits associated with soil investments and thus encompass a major portion of the potential benefits from the farmers' perspective.

Soil Resources and Average Agricultural Productivity Levels

Lal has identified the following processes which are included in soil degradation: accelerated erosion and desertification, compaction and hard setting, acidification, decline in soil organic matter content and biodiversity, and depletion in soil fertility. These processes result in changes in soils which in turn have an impact on yields through changes in soil nutrient content, water-holding capacity (WHC), organic matter content (SOM), soil reactivity, topsoil depth, salinity and biomass.

Soil degradation effects on agricultural productivity are manifested through their impacts on both the average and variance of yield, as well as the total factor productivity of agricultural production. These impacts are translated into economic costs in the form of loss of income (or consumption), increased income (or consumption) risk, and increased costs of production. Several studies have been conducted in an attempt to quantify these costs under varying circumstances. Their results have indicated that these can be quite substantial.

The impact of soil degradation on agricultural productivity is a dynamic process that is highly heterogeneous across time and space. In addition, the impacts vary by type of degradation: the impact of soil erosion on agricultural yields is different than that of salinisation. The fact that soil degradation impacts are so highly site and time specific is one reason that it is difficult to assess the economic implications in a generalised fashion. In addition, the fact that humans can influence both the occurrence and impacts of various forms of degradation also makes it difficult to assess the economic implications of the process.

To illustrate the impact of land degradation on agricultural yields, the case of soil erosion can be taken. Soil erosion is caused by water and wind action leading to a loss of soil from the uppermost land layers. Oldeman, Hakkeling and Sombroek estimated that approximately 83% of the land degradation in developing countries is caused by water and wind induced soil erosion. Erosion can result in the physical deterioration of the soil and/or induce chemical deficiencies and toxicities, which in turn impact agricultural crop yields.

Yadav and Scherr summarise over 40 studies at a global, regional and local scale of analysis on the impacts of erosion on agricultural production. The estimates of impacts are reported in terms of yield declines as well as economic values. Yield declines range from 0.04 percent per annum to 100 percent (e.g. land goes out of production). In terms of economic costs, estimates range from US$26 000 million annually on a global scale to 5-10 percemt of GNP for developing countries annually. At a farmer level, estimates of losses of between 0.4 to 8 percent of agricultural gross product have been made. Several different methodologies and assumptions are used in the calculation of both the yield and economic impacts of degradation among these studies, so their results are not directly comparable.

Erosivity and Erodibility

The relationship between soil erosion and agricultural yields can be divided into two separate components: the rate at which erosion occurs over time, e.g. the susceptibility of the soil to erosion as measured by tonnes of soil loss per year, and the relationship between the level of erosion and productivity measured in tonnes of crop yield loss per unit erosion. The first is a function of two groups of factors: forces that generate erosion and forces that resist erosion. Forces that generate erosion include climate and topography; specifically the amount and distribution of rainfall and wind velocity and the steepness and length of slope. Collectively these are referred to as the erosivity of the soil. The forces that resist erosion include the degree of land cover as well as several characteristics of the soil including its texture, structure, water retention capacity and transmission properties. Taken together this latter group of characteristics may be referred to as the *erodibility* or resilience of the soil.

In the tropics there is some evidence that soils exhibit high rates of erosivity and low rates of erodibility. One major factor is that rainfall patterns in the tropics tend to have sharp and intensive peaks in distribution that is highly erosive. Tropical soils are generally considered less resilient to erosion, because they are more highly weathered. For example, Alfisols that are widespread in arid and semi-arid areas of South Asia and sub-Saharan Africa have extremely poor soil structure and are highly prone to accelerated erosion . However, Lal makes clear that it is not possible to generalise that in the tropics there is a higher rate of erosion occurring over time than is the case in temperate regions.

The negative relationship between the level of erosion and impact on crop yields is generally expected to be stronger in the tropics than in temperate zones. Soil erosion causes more severe quality changes in tropical soils due to their low inherent fertility, concentration of plant available nutrients or organic matter in the topmost layers of the soil, and subsoils unfavourable for edaphic life forms and lacking effective root volume. However, as Sanchez and Logan point out, it is important to disaggregate specific elements of tropical soil characteristics in order to illustrate the distribution of soil chemical properties relating to fragility throughout the tropics, rather than simply assume tropical soils are universally infertile and fragile.

Extent of Erosion

The degree to which the soil has already experienced human-induced or natural degradation is clearly an important determinant of the current yield impacts of degradation, as well as potential returns to reversing or continuing along the path of erosion. That is, one needs to know not only what the shape of the function mapping erosion over time looks like, but at which point on the function the farmer currently finds himself or herself. There is a fairly high probability that a land user will be operating on degraded land. In the Global Assessment of Soil Degradation study, Oldeman, Hakkeling and Sombroek found that 23 percent of all lands in agriculture, pasture or forestry were degraded to some extent.

The costs of erosion to a farmer in terms of foregone output, then, depend upon the following four factors: (i) the resilience of the soil; (ii) the degree of erosion risk present (e.g. the erosivity); (iii) the degree to which erosion levels impact yields (the sensitivity); and (iv) the degree to which erosion and land degradation are already present. In essence, one needs to know the shape of the functions relating yields to level of erosion as well as erosion over time and the point at which the farmer or land user finds himself/herself on this latter curve, in order to determine the costs of land degradation and potential benefits from reversal to the farmer.

Considerable research on the rate of erosion over time on varying soil types and under varying production conditions has been done. In general, with no technological change intervening, the function is non-linear and convex to the origin: erosion increases over time at an increasing rate.

Relationship between Erosion and Yields

Research on the nature of the relationship between erosion levels and yields is less common because of difficulty in measurement due to the large number of variables involved as well as the interactions between them. Recent research by FAO has indicated that for a wide variety of soil types, there is a negative exponential, or logarithmic relationship between level of erosion and yields. The implications of this finding are that the highest returns in terms of crop yields to reversing or decreasing soil erosion will be realised on less-eroded soils. This relationship was found to be fairly robust for the humid and subhumid tropics, but less so for semi-arid areas, where water availability becomes the main constraint to productivity.

Erosion has an impact on crop yields through changes in soil chemical characteristics and/or changes in soil physical or structural characteristics. One of the most widespread examples of the former is the loss of soil nutrients through leaching associated with erosion. The highest concentration of nutrients necessary for plant growth is found in the uppermost layers of the soil, so the loss of this layer results in a decrease in soil nutrient reserve. The situation is even more serious on soils with low levels of natural fertility, as is the case in much of the tropics.

Erosion also causes the loss of a buffer layer of organic material, exposing soils to aluminium toxicity and acidification which can cause sudden and severe yield decreases. Through the removal of clay content and organic matter, erosion may result in a decreased capacity of the soil to provide phosphorus in a form usable to plants (e.g. increase phosphorus fixation). In terms of structural impacts, soil erosion can increase the bulk density of the soil, making it more difficult for water to penetrate to rooting depths and for plant shoots to emerge, either by the removal or organic matter and colloids that create spaces between soil particles or by exposing highly compacted subsurface layers.

Farmers can mitigate the chemical or physical consequences of erosion by substituting for the services lost through the erosion process, or correcting for the chemical and physical constraints erosion leads to. The application of fertilisers to replenish lost nutrients is the most common example of substitution. Correcting acidity through the addition of lime is an example of a corrective action. However, soil natural capital provides a wide array of services in conjunction with other components of the ecosystems it is embedded within, so even if substitution for one aspect of the services provided by soil capital is attainable, there may still be a decrease in other ecosystem services. The decline in these linked services may result in a decrease in the productivity of manufactured inputs, with the result that under some circumstances, soil natural capital and human applied manufactured capital are complements, rather than substitutes. In the case of chemical fertilisers, for example, the loss of soil quality through degradation processes may result in a decline in the yield response to chemical fertilisers, e.g. the marginal productivity of chemical fertiliser declines with declining soil natural capital.

The question of the extent to which manufactured forms of capital can substitute for natural capital is quite controversial and lies at the heart of the debate on sustainable natural resource management. Weak sustainability is defined as the maintenance of an aggregate level of the capital stock, be it composed of natural or human-made capital. In contrast, strong sustainability requires the maintenance of natural capital stock. The difference between the two positions is essentially a difference in the assumption of substitutability between natural and manufactured forms of capital. Proponents of weak sustainability assume a high degree, while those of strong sustainability think of natural and manufactured capital as complements.

Whether or not soil natural capital and manufactured capital are substitutes or complements is a critical determinant of the impacts of soil erosion on productivity and thus farm decision-making. Essentially, if they are complements then the loss of soil natural capital is higher, because it results in both a loss in productivity directly associated with its own loss, as well as a decrease in productivity in manufactured inputs. Deriving parameters that describe under which conditions natural capital and manufactured capital are substitutes or complements is an important area for future research.

In several economic models of optimal soil management the results are dependent upon the critical assumption of complementarity between natural and manufactured inputs. Goetz constructed a dynamic model of soil management in which he distinguished between the case of soil depth and fertiliser as substitutes and one as complements. He hypothesized that the two would be complements at low levels of soil depth where depth is the limiting factor, but substitutes at high levels of soil depth due to a lower marginal accretion in the yield as the soil increases. The results of his analysis indicate that which of these cases holds true is a critical determinant of the steady state soil stock, as well as the impact of changes in crop prices or the discount rate on the optimal level of soil depth. In his dynamic models of optimal soil management, Clarke assumes that soil natural capital and manufactured inputs are complements but notes that "the relationship between commodity price changes and viable soil conservation strategies depends crucially on the complementarity/substitutability relationships between the inputs". La France and Barbier assume a complementary relationship between soil stock and cultivation and Grepperud side-steps the question by defining the inputs to the production function in relation to their impact on soil quality or erosion, rather than including an explicit term for soil natural capital in the production function.

In an empirical study, Walker and Young found that, with fertiliser inputs, higher yields were obtained on fields with lower levels of erosion, indicating a complementary relationship between soil natural capital and chemical fertiliser. They analysed the impacts of erosion on crop yields in the Palouse area in Idaho and Washington States in the United States that had experienced both high rates of erosion and increasing yields. The authors concluded that the complementary relationship between inputs is highest in situations where unfavourable subsoils underlie topsoil horizons. A greater degree of substitution would be found between chemical fertilisers and eroded soils in cases where a greater degree of homogeneity exists between surface and subsoil quality.

Since technological change itself is a complement to soil natural capital, any assessment of the impacts of soil quality decline must consider the difference between yields obtainable under the new technology with and without the presence of erosion, rather than yields before and after the adoption of the technology, which is often the case among farmers. According to the authors of the study, soil conservation practices were not widely adopted in the area because of the producer's belief that yield-enhancing technology compensated for lost topsoil.

Impact of Erosion on Crop Yield Variability

Another major concern of farmers is the variability of crop yields over time. Yield fluctuations arise primarily due to climatic factors or pest and disease incidence. Fluctuations in yields create production risk, which generates costs to risk averse farmers seeking to insure themselves. The absence of facilities to store produce over time, together with poor market development and low incomes, means that farmers are often directly dependent on the current year's harvest for their food supply. Variations in yields can have quite dire consequences for rural populations and therefore yield variability enters production decision-making as a source of risk, in some cases quite intensely.

One of the biggest sources of variability in yields is uncertain rainfall. The inability of crops to obtain sufficient water during critical periods in the growth process results in significantly lowered or no yields at all. In much of the world, most farmers are dependent on rainfall for their source of crop water. Water is made available to plants through transport through the soil medium into the plant roots. The capacity of the soil to store water that can be made available to the plant over the growing season is thus an important feature for mitigating uncertain rainfall patterns. This capacity, known as the soil water holding capacity (WHC). Under undegraded conditions, soils vary naturally in their WHC, with heavier soils with higher clay content exhibiting higher levels generally than lighter, sandy soils.

The process of degradation can reduce the soils' WHC. The loss of soil topsoil layers through erosion may have a negative impact on soil moisture retention capacity and increased infiltration rates. Colacicco, Osborn and Alt state that soil erosion may increase the variability of production regardless of its effect on average yields. Moyo found that soil erosion in Zimbabwe resulted in reduced soil organic matter and clay content, which caused a drop in fertility, an increase in bulk density and a reduced storage capacity of plant available water, particularly with the sandy soils in Zimbabwe's communal areas.

His research was carried out in the semi-arid region of southern Zimbabwe that is characterised by erratic and unreliable rainfall between and within season. The results indicated that there were no significant yield differences in years of high rainfall, but under poor rainfall conditions yields varied significantly, with plots treated with mulch ripping showing much higher yields in the poor rainfall year.

To what extent can farmers substitute manufactured inputs to substitute for the lost services of the soil in terms of yield variability? In the case of rainfall uncertainty, one obvious measure that could be taken is the supply of irrigation water to substitute for lost water availability in the soil. However, as in the case of declining average yield levels, the adoption of such mitigating measures in the presence of continuing soil degradation results in a decreased productivity of irrigation water—for example, soil quality and irrigation are complements rather than substitutes. Under degradation that leads to decreased infiltration rates and lower moisture retention rates, higher and more frequent irrigation rates are required to obtain a given level of water delivery to a plant. Under degradation, irrigation efficiencies (e.g. the degree of water used by the plant divided by the total amount of water applied) decline, thus the costs of irrigation are higher. If the fixed capital costs of constructing and maintaining an irrigation system are considered as well as the variable costs of delivering the water each season, the returns to using irrigation as a substitute for services lost through soil degradation are even lower.

The implication of these findings is that the cost of soil degradation to farmers is higher than just loss in average yields, particularly in rainfed areas that experience uncertain rainfall patterns. Under these conditions, investments in soil conservation which result in reduced yield variability may be quite attractive to farmers, even if they do not increase average yields. The critical determinants of the decision to adopt soil conservation measures in this case are the degree of exposure to risk arising from the degradation and the farmer's attitude towards risk on the one hand, as well as the relative costs and returns to soil conservation and technologies which mitigate the impacts of yield variability.

Costs of Soil Conserving and Mitigating Inputs

In the previous two sections the major impacts of soil degradation, and specifically erosion, on farm production costs and the strategies which the farmer can adopt to address these were outlined. Broadly, farmers can opt for one of four strategies: (i) do nothing; (ii) take measures to slow the rate of erosion; (iii) take measures to mitigate the negative impacts of erosion on crop yields and variability; and (iv) adopt a combination of (ii) and (iii). The strategy the farmer adopts will depend on the magnitude and type of costs associated with degradation, and how these compare with the costs associated with adopting any one of these four strategies.

A key concept necessary to understanding the costs of soil management strategies from the farm perspective is that of opportunity costs, which is simply the value foregone by employing a resource in one use, rather than an alternative use. This means consideration of what the farmer is giving up in adopting a particular course of action, which may involve financial costs such as payments for inputs or loss of livestock feed supplies, as well as non-market costs, such as the loss of leisure or a loss of insurance capacity. As financial constraints are a key barrier to the adoption of any farm management strategy, it is essential to consider the strict financial implications associated with any course of action; however, in order to have a full understanding of the incentives and constraints of the farmer in making investment decisions on soil natural capital, it is necessary to consider other non-financial costs as well.

Taking each of the key factors of production in turn, a broad description of the types of opportunity costs associated with each, if they were to be employed in any one of the four strategies outlined above, can be outlined. For example, in the case of the first strategy, the cost to the farmer of doing nothing in the face of soil degradation associated with erosion is the loss of production value associated with the degradation—the loss of yields, input efficiency or crop yield stability which is due to the degradation. The magnitude of this loss will vary considerably, depending on the rate of erosion, the rate of yield loss or stability loss associated with erosion, and the current level of erosion. In some cases this loss may be rather small, and less than the costs to the farmer of adopting any measures to reverse the process, in which case the farmer has no incentive to take any action; the opportunity cost of the foregone production is less than the cost of preventing its loss.

Now take the case where erosion does result in substantial production costs and assume that the costs associated with slowing or reversing the erosion process are less than the foregone production losses. Ostensibly the farmer has a financial incentive to adopt erosion control. However, if the erosion control measure involves the use of investment funds in the current time frame in order to achieve a positive impact on production in a future time frame, the farmer is essentially giving up the use of this capital in the present—foregoing the current use of capital. The way in which this trade-off of current capital for future capital is viewed depends on the discount rate of the farmer, as well as the expected future payoff. A high discount rate, which implies that current consumption is weighted more heavily than future consumption, will result in a high opportunity cost of investment funds to the farmer.

Another type of opportunity cost that a fixed capital investment for erosion control might involve is that of insurance capacity. By making an "irreversible" capital investment in erosion control, the liquidity of the asset portfolio of the farmer is lowered. To the extent that asset liquidity serves as an insurance mechanism that allows for a rapid access to funds in the case of shocks, tying up funds in an illiquid investment implies foregoing an insurance benefit.

The opportunity costs associated with labour are also a critical determinant of farmers' soil management strategies. Adopting erosion control or mitigation measures may require more labour, or a more arduous type of labour, or a labour requirement at a specific seasonal time. In adopting such a measure the farmer may be giving up the opportunity to work off-farm—and thus the wages associated with such employment. Alternatively, the farmer may be foregoing leisure, or less physically demanding types of labour, or the ability to work on other crop activities on farm. Any of these may constitute a significant cost to the farmer, and thus a disincentive to adopt the strategy in question. Under these conditions, farmers are unlikely to invest at a socially optimal level in avoiding soil erosion.

The opportunity costs of other forms of natural capital are another key consideration determining the incentives of farmers to invest in soil natural capital. That is, if natural capital is employed for the purposes of preserving or augmenting soil natural capital, what alternative uses are foregone? One important example in this category is the use of crop residues as a mechanism for returning biomass to the soil, as well as providing erosion-reducing groundcover. Crop residues may be used for livestock feed or fuel purposes; however, in this case the loss of these uses constitutes a significant opportunity cost to the farmer, particularly where alternative sources of fuel and feed are scarce and expensive.

Given this framework for assessing the costs associated with the various options for soil management, an assessment can be made of the empirical literature on experiences with the adoption of soil erosion and mitigation practices on farm. Traditionally, measures designed to decrease soil erosion, have been primarily focused on engineering approaches, for example the construction of physical barriers such as terraces or bunds to impede the movement of the soil. Such measures tend to require a large fixed capital investment and/or a high input of labour. The impacts or benefits of these barriers tend to be realised only after a substantial length of time. In addition, in many cases the adoption of such technologies leads to a decrease in crop output, due to a loss of area in production. The combination of a high fixed cost up front and delayed and reduced benefits make this sort of investment unattractive in many cases to farmers.

Lutz, Pagiola and Reiche reviewed cost benefit analyses for nine soil conservation projects in Central America and found that the measures with the highest economic return were the low cost and quick returning ones, in areas where erosion had appreciabe impacts on productivity. In one case, the adoption of soil conservation led to a decline in long-term returns to agriculture, due to the loss of land in production to diversion ditches and the limited impacts of erosion on productivity. Barbier found that the high labour cost for the construction of bench terraces to stop erosion in the uplands of Java made the adoption of this measure unattractive to farmers. This implied a loss of foregone income from the labour that could be used in alternative income-generation activities, as well as a loss of land for production.

In Honduras, Carcamo *et al.* distinguished between the on-farm versus externality benefits of adopting various measures that impact soil degradation rates, growing various crops with varying land cover services, the use of alternative tillage practices, and the construction of soil erosion control devices. They concluded that in order to achieve socially optimal levels of soil degradation both a reduction in farmer income and an increase in the degree of production risk the farmer is exposed to will occur, which implies a need for subsidies or transfer payments to farmers in order to achieve the desired level of degradation control.

Hwang *et al.* conducted research on farmer incentives for erosion control among small-holders on steeply sloping lands in the Dominican Republic and found that large-scale reduction in soil loss can only be achieved at the expense of significant declines in farm income. Even where soil erosion control measures result in increased production, the impact of erosion control adoption on production risk may make them unattractive to farmers.

More recently attention has become focused on biological measures of soil erosion control, such as cover crops and mulching. The attractiveness of these measures as compared with engineering approaches are a lower initial investment requirement, less loss of cropping area, and the realisation of benefits other than just erosion control, such as fertility enhancement, soil structure improvement and weed control. However, they too involve costs to the farmer. Erenstein gives great detail on the costs which may be involved with the adoption of crop residue mulching over conventional systems, including increased seeding, fertilisation and weeding costs, losses in input productivity and loss of livestock grazing. He also makes the point that even in cases when adoption of a mulching technology results in an overall reduction in labour requirement, the timing of the labour input may shift and conflict with other activities and therefore not be a viable option for farmers.

The costs of adopting substitution and correction approaches to erosion-induced productivity loss are those of obtaining and applying the inputs. In general these types of measures tend to fall into the category of operating rather than fixed costs—e.g. they occur on an annual or seasonal basis and their impacts are generally realised within the same production cycle. For this reason, as well as their generally lower costs as a percentage of overall production costs, mitigation measures are often more attractive to farmers than erosion control measures. They do not involve a high rate of foregone current use of capital, they allow some flexibility in terms of ability to respond to shocks, and they exhibit short-term and visible results. The problem is, of course, that the effectiveness of mitigation strategies will decline with increasing soil erosion, although this is not always recognised by the farmer.

REFERENCES

Barbier, E.B. 1989. *Economics, natural-resource scarcity and development: conventional and alternative views.* London, Earthscan.

Bebbington, A. & Perreault, T. 1999. Social capital, development, and access to resources in Highland Ecuador. *Economic Geography*, 75(4): 395-418.

Binswanger, H. 1990. Attitudes toward risk: experimental measurement in rural India. *American J. Agricultural Economics*, 90: 195-407.

Bishop, J. & Allen, J. 1989. The on-site costs of soil erosion in Mali. *World Bank Environment Working Paper No. 21.* Washington DC

Blaikie, P. & Brookfield, H., eds. 1987. *Land degradation and society*. London and New York, Methuen.

Bromley, D. 1998. Property regimes in economic development: lessons and policy implications. *In* E. Lutz, ed. *Agriculture and the environment: perspectives on sustainable rural development.* Washington, DC: World Bank

Cuesta, M., Carlson, G. & Lutz, E. 1997. *An empirical assessment of farmers' discount rates in Costa Rica.* Environment Department, World Bank: Washington, DC.

Dasgupta, P. & Maler, K.G. 1995. Poverty, institutions and the environmental resource base. *In* J. Behrman and T.N. Srinivisan, eds. *Handbook of development economics*, Vol. III. p. 2371-463. Amsterdam, Elsevier.

Dixit, A. & Pindyck, R. 1994. *Investment under uncertainty*. Princeton, USA, Princeton University Press.

FAO. 1978. *Report on the Agro-ecological Zones Project. Vol. 1: Methodology and results for Africa.* World Soil Resources Report No. 48. Rome.

Grepperud, S. 1997a. *Soil depletion choices under production and price uncertainty*. Discussion Paper 186. Oslo, Statistics Norway

Bibliography

Atkinson, H.J. 1973: The Respiratory Physiology of the Marine Nematodes *Enoplus brevis* (Bastian) and *E. communis* (Bastian): I. The Influence of Oxygen Tension and Body Size. *J. Exp. Biol.* **59**(1): 255-266. PDF fulltext

Bell, G 1982: *The Masterpiece of Nature: The Evolution and Genetics of Sexuality*. University of California Press.

Bennett, Gary W., Ph.d., Owens, John M., Ph.d., Corrigan, Robert M, Ph.d. *Truman's Scientific Guide to Pest Management Operations*, 6th Edition, pages 10, 11, 12, Purdue University, Questex, 2005.

Berry, R. E. 1972. Garden symphylan: Reproduction and development in the laboratory. (*Scutigerella immaculata*). Journal of Economic Entomology. Vol. 65. p. 1628-1632.

Bollen, Walter B. 1959. *Microorganisms and Soil Fertility*. Oregon State College. Oregon State Monographs, Studies in Bacteriology, No. 1. 22 p.

Boyle, M., W.T. Frankenberger, Jr., and L.H. Stolzy. 1989. "The influence of organic matter on soil aggregation and water infiltration". *Journal of Production Agriculture*. Vol. 2. p. 209-299.

Campbell, Robert N., and Arthur S. Greathead. 1990. "Control of clubroot of crucifers by liming". p. 90-101. In: A.W. Engelhard (ed.). *Management of Diseases with Macro- and Microelements*. APS Press. American Phytopathological Society, St. Paul, MN.

Ciesla, W.M. 1994. *Assessment of the 1994 outbreak of nun moth, Lymantria monacha in Poland and proposed pest management activities*. FAO, Rome, Italy FO 2/354, 24 pp.

Comis, Don. 1997. "Glomalin-soil's superglue". *Agricultural Research*. USDA-ARS. October. p. 22.

Cook, R.J., and K.F. Baker. 1983. *The Nature and Practice of Biological Control of Plant Pathogens*. APS Press, St. Paul, MN.

Dick, J.B., and H.B. Tisdale. 1938." Fertilizers in relation to incidence of wilt as affecting a resistant and susceptible variety". *Phytopathology*. Volume 28. p. 666-667. (abstract).

Edwards, C. A. 1957. Simple techniques for rearing Collembola, Symphyla and other small soil-inhabiting arthropods, pp. 412-416. In D. K. M. Kevan [ed.], Soil Zoology. Butterworths Publications Ltd., London.

Engelhard A.W. (ed.). *Soilborne Plant Pathogens: Management of Diseases with Macro- and Microelements*. American Phytopathological Society. 217 p.

Filinger, G. A. 1928. Observations on the habits and control of the garden centipede, *Scutigerella immaculata*, Newport, a pest in greenhouses. J. Econ. Entomol. 2: 357-360.

Getzin, L. W., and C. H. Shanks. 1964. Infection of the garden symphylan, *Scutigerella immaculata* (Newport), by *Entomophthora coronata* (Constantin) Kevorkian and *Metarrhizium anisopliae* (Metchnikoff) Sorokin. J. Insect Path. 6: 542-543.

Goldstein, Jerome. 1998. "Compost suppresses disease in the lab and on the fields". *BioCycle*. November. p. 62-64.

Gould, G. E., and C. A. Edwards. 1968. Damage to field corn by symphylans. Proc. Indiana Acad. Sci. 77: 214-221.

Granatstein, David. 1998. "Suppressing plant diseases with compost". *Good Fruit Grower.* May 1. p. 9-11.

Howitt, A. J. 1959. Control of *Scutigerella immaculata* (Newport) in the Pacific Northwest with soil fumigants. J. Econ. Entomol. 52: 678-683.

Hudson, Berman D. 1994. "Soil organic matter and available water capacity". *Journal of Soil and Water Conservation.* March-April. p. 189-194.

Illingworth, J. F. 1927. Symphylids destructive to the roots of pineapple. Pineapple News Mar-Dec: 88-91.

Ingham, Elaine R. 1991. "Interactions among mycorrhizal fungi, rhizosphere organisms, and plants". p. 169-197. In: P. Barbosa, V.A. Krischik, and C.G. Jones (eds.). *Microbial Mediation of Plant-Herbivore Interactions.* John Wiley and Sons, Inc. 530 p.

Jackson, William R. 1993. *Organic Soil Conditioning.* Jackson Research Center, Evergreen, CO. 957 p.

Jahn, GC, JA Litsinger, Y Chen and A Barrion. 2007. *Integrated Pest Management of Rice: Ecological Concepts.* In Ecologically Based Integrated Pest Management (eds. O. Koul and G.W. Cuperus). CAB International Pp. 315-366.

Johnston, S.A., and P.J. Nitzche. "Rotation periods suggested to help control vegetable diseases". New Jersey Extension Service. 1 p.

Kogan, M 1998. "Integrated Pest Management:Historical Perspectives and Contemporary Developments", *Annual Review of Entomology* Vol. 43: 243-270.

Koontz, F. R. 1968. Biological and ecological relationships of the fungus, *Entomophthora coronata* (Constantin) Kevorkian, and the garden symphylan, *Scutigerella immaculata* (Newport). Ph.D. dissertation, Oregon State University, Corvallis.

Logsdon, Gene. 1995. *Using compost for plant disease control. Farm Scale Composting.* JG Press, Inc. Emmaus, PA.

Matsubara, Yoh-ichi, Haruto Tamura, and Takashi Harada. 1995. "Growth enhancement and verticillium wilt control by vesicular-arbuscular mycorrhizal fungus inoculation in eggplant". *Journal of Japanese Horticultural Society.* Volume 64, Number 3. p. 555-561.

Matsubara, Yoh-ichi, Haruto Tamura, and Takashi Harada. 1995. "Growth enhancement and verticillium wilt control by vesicular-arbuscular mycorrhizal fungus inoculation in eggplant". *Journal of Japanese Horticultural Society.* Volume 64, Number 3. p. 555-561.

Nelson, E.B., L.L. Burpee, and M.B. Lawton. 1994. "Biological control of turfgrass diseases". p. 409-427. In: A.R. Leslie (ed.). *Handbook of Integrated Pest Management for Turf and Ornamentals.* Lewis Publishers, Ann Arbor, MI.

Newhook, F.J. & Podger, F.D. 1972. The role of *Phytophthora cinnamomi* in Australian and New Zealand Forests. *Annual Review of Phytopathology* 10: 299-320.

Newsham, K.K., A.H. Fitter, and A.R. Watkinson. 1995. "Arbuscular mycorrhiza protect an annual grass from root pathogenic fungi in the field". *Journal of Ecology.* Volume 83. p. 991-1000.

Obiri, J.F. 1994. Variation of cypress aphid, *Cinara cupressi* (Buckton), attack on the family Cupressaceae. *Commonwealth Forestry Review* 73: 43-46.

Ozores-Hampton, Monica, H. Bryan, and R. McMillian Jr. 1994. "Suppressing disease in field crops". *Bio-Cycle.* July. p. 60-61.

Sances, Frank V., and Elaine R. Ingham. 1997. "Conventional organic alternatives to methyl bromide on California strawberries". *Compost Science and Utilization.* Spring. p. 23-37.

Schneider, R.W. (ed.). 1982. *Suppressive Soils and Plant Disease.* The American Phytopathological Society. St. Paul, MN. 88 p.

Index